Farming and Subsidies

Brian Chamberlin

To Jan, who fought the battles with me;
and to the farmers of the future, for whom they were fought.

First published in 1996 by Euroa Farms Ltd (Publishing Division),
PO Box 324, Pukekohe, New Zealand.

ISBN 1-86956-153-8

Research/editing: Jan Chamberlin
Copy editing: Linda Pears
Design: Todd Harding
Cover photograph: Andris Apse

Printed by GP Print, 10 Mulgrave St, Wellington, New Zealand.

CONTENTS

INTRODUCTION

My main purpose in writing this book is to show that it is possible to dismantle an agricultural subsidy system without causing all the disasters that subsidy supporters predict.

Myths have sprung up about the terrible harm that subsidy removal will do to the farmers, the farm service industries, the environment, land and stock prices, employment, food security and food safety.

New Zealand has a special place in dispelling these myths. We built up a subsidy system and then removed it. The New Zealand experience has shown that many of the dire predictions did not happen - and that they are unlikely to happen in other countries that follow New Zealand's lead.

The book begins by giving some background to the beginning of subsidies in New Zealand, showing the distortions they caused and the distortions they are still creating in other countries. It follows the process of reform, indicating where the process could have been improved, and how the farming sector coped with the huge changes it had to go through.

Other major issues discussed are the fight for fair trade through the Uruguay Round of the General Agreement on Tariffs and Trade, the cost of subsidies to consumers and taxpayers, and why subsidies do not help to feed the hungry.

It is not my intention to provide an academic or theoretical discussion of the arguments about subsidies and fair trade that have raged in world farming circles over the last decade or so, but to give an inside view - the view of someone who was there as it all happened.

Along with other farmers, my family suffered under the effects of high inflation, subsidy removal, high interest rates and low incomes. This meant that my opinions and actions while I was leading Federated Farmers at the national level impacted on my family's livelihood; I also had to convince farmers that what we were doing was right.

Internationally, I have fought for fairer trade, and for New Zealand's case, in farming organisations and in political and diplomatic circles. I was in Brussels when the European farmers marched in the streets and created havoc to convince people that they should hold on to their privileged position. I have been to some of the poorest countries in the world and watched the people trying desperately to compete against unfair trade practices.

Having been deeply involved in this era as a farmer, farm leader (national and international), lobbyist, diplomat and negotiator, I felt that I should try to share the convictions that I have come to as a result of my experience. In particular, I hope to be able to show how it really was in New Zealand, especially on the farms; to contribute to dispelling some of the myths that have grown up around the removal of subsidies; and to try to correct some of the misinformation being put about by the opponents of reform, even today.

A FOOL'S PARADISE

New Zealand was, for many generations, regarded as Britain's farm and food supply. Many of the people who immigrated here from Britain and Ireland in the second half of the nineteenth century came because it was possible for them to buy land and farm it on their own account. This would have remained an impossible dream if they had stayed in the countries where they were born.

Even those who came here to look for gold, or as servants to the wealthy immigrants and landowners, soon turned to farming when they found that it was relatively simple to buy their own land.

Farming was certainly not easy for most of these people. The land had to be cleared of native bush, burned, seeded and grassed. It also had to be fenced and stocked. The cows had to be milked and the sheep shorn in very primitive conditions. Roads were poor, and it took many years before things improved.

Initially, much of the farming could be described as subsistence. Cattle, sheep, pigs and poultry were kept in tiny numbers to supply family needs. The cows provided milk for drinking and cream for homemade butter. The meat was salted down and the eggs preserved for the winter. Wheat was grown for flour, and oats for the horses, which provided the transport and the horsepower for ploughing. Vegetables, especially potatoes, were vital for the family's survival.

If farmers had a surplus, they sold it locally. Only non-perishables could be shipped on long voyages, so wool was the major export earner. Along with other non-perishable goods it was sold to Britain, which gladly bought all New Zealand produced.

Inevitably there were serious setbacks but there were also great developments, the major one being refrigerated shipping, which meant that it was possible to market a much greater variety of produce, particularly meat and dairy products, around the world. This led to much more specialisation in both dairy and sheep farming.

An early setback was that the initial fertility of the land soon ran out, and some of it had to be abandoned. But the introduction of fertiliser solved the problem. The fertiliser industry grew, and farming grew with it. This process was further boosted when pilots returning from World War II pioneered the spreading of fertiliser by air in the early 1950s. Many thousands of hectares of previously unproductive hill country were eventually made useable and fertile in this way.

The adage 'New Zealand rides on the sheep's back' was often quoted, and was reasonably true; if cows had been included, it would have been totally accurate. This was because for many years almost all of the country's overseas funds came from the farming industry. Most employment was related to farming, directly or indirectly.

Although most of the produce was sold in Britain, prices fluctuated with the volatile world market. Everyone in New Zealand prospered when prices were good, and everyone suffered in times of recession. The depression of the 1930s was felt by all sectors of the community. In 1935 a Labour Government swept into power on a promise to eliminate poverty. It was thought that policies could be introduced that would insulate New Zealand from fluctuations in commodity prices.

This government did indeed introduce many new ideas. For example, it created credit to carry out a large programme of public works. Housing was a priority, and many houses were built using this finance, which was lent at very low interest rates. A manufacturing industry was established through the introduction of a licensing system that made it very difficult to import goods into the country.

The 'welfare state' was greatly expanded, with the payment of social security benefits to young and old and others in need, and the provision of health services and education free for all New Zealanders.

Farming was also involved in the new policies, with a guaranteed price being set up for dairy farming. Reserve Bank credit at an interest rate of 1 percent per year was made available to the dairy industry, for farmers to use as a cushion against a market price collapse.

For the next twenty years these policies were never really tested. World prices for agricultural products were recovering by the time the policies were implemented, and right through World War II and the period that followed New Zealand was able to sell all it produced at favourable prices.

The commodity booms associated with the Korean War and the Suez crisis put many millions of pounds into farmers' pockets, and consequently into the New Zealand economy during the 1950s. The country prospered, and at one point it was the second wealthiest in the world as measured by income per head of population. The sheep farmers did so well from the Korean War wool boom that the government actually had to freeze a portion of their income to dampen down the economy.

When I look back to my youth, I am reminded how good things were. My parents, for example, owned a small sheep and cattle farm on which they ran 800 ewes and a few cattle. In the early fifties, this 175-acre (70-hectare) farm provided them with enough income to own a cottage at the seaside, a launch and a Plymouth Savoy car. They were also able to send me and my brother to boarding school.

My parents were very hard-working. The beach cottage was small and the boat was no 'gin palace'. However, no farm of that size could support the same lifestyle today. Many thousands of sheep and hundreds of cattle would now be needed to provide a similar standard of living.

The wealth was reasonably evenly spread throughout the community. There was full employment, and a central wage-fixing system increased wages by at least the rate of inflation. If prices went up, wages were increased, so the population was insulated from any fundamental changes that might have been taking place.

Compulsory unionism gave the unions huge power. They often called strikes to secure wage increases that went beyond the rate of inflation. Vulnerable seasonal industries such as those involved in processing perishable farm produce were soft targets and consequently paid dearly.

Which party got elected to government largely depended on which promised the most. The National Party defeated Labour in 1949, and stayed in government until 1957, when Labour won again after promising taxpayers a tax refund of £NZ100 each . Their Finance Minister, Arnold Nordmeyer, then introduced a tough budget to pay for the refund and to deal with an increasingly difficult financial situation.

What became known as 'Nordy's Black Budget' not only led to Labour's demise in 1960, but also did more than anything else to keep National in power for the next twelve years. National did not change any of the policies Labour had introduced. If anything, society became even more regulated and protected. Just about everything was licensed or centrally controlled. It seemed as if all that was required for National to win each election was for someone to mention that infamous 'Black Budget'.

If we compare Federated Farmers' concerns and policies of today with those of many years ago, we find that they are amazingly similar. For example, the organisation has always opposed import controls. It seems hard to credit, therefore, that the farmer-dominated National Party of the fifties, sixties and seventies defended the implementation of import controls, especially when a number of the cabinet ministers were former officeholders in Federated Farmers, and the Prime Minister of many years, the Right Hon. Sir Keith Holyoake, was one of the organisation's founding members. In retrospect, I think it can be said that for some years the farming situation was so good that the industry was able to carry the cost of such impositions, and the will to fight them probably went out the window.

As technology developed, the rural population declined and the cities grew. Although many politicians came from the country, it was the urban people who had the numbers when it came to elections and who were perceived to gain from the protection of manufacturing industries. Getting elected was always the priority for most politicians and National, even with the bogey of the 'Black Budget', may not have survived a dismantling of import licensing.

So life went on in this protected and controlled little country, and no-one realised that the world was changing and that even in those days we could not really afford the lifestyle we enjoyed. In our innocence, we assumed the good times would go on forever.

A fool's paradise indeed.

THE REALITY

The golden days of the 1950s were followed by much more difficult times for New Zealand, and for its farming industry in particular.

Great changes were taking place in Europe as the population picked itself up from the ravages of war. People who had almost starved through two world wars were no longer prepared to depend on their various colonies for a large proportion of their food. They were happy to go along with the notion of subsidising their farmers to encourage them to produce to a level of self-sufficiency.

In 1958 Belgium, France, West Germany, Italy, Luxembourg and the Netherlands formed a Common Market. Central to this was an agricultural policy which subsidised farmers' incomes and protected them from import competition. So central was it, in fact, that over the years the Common Agricultural Policy (CAP) used up to 80 percent of the budget of the Common Market (or European Economic Community (EEC) or European Union (EU), as it successively became known).

Dairy produce, sheep meat and wool at this time accounted for most of New Zealand's exports. Britain was the main recipient, although some of the wool was bought by mills in continental Europe. New Zealand was extremely vulnerable because it was exporting a narrow range of goods to what was virtually a single market.

Technological developments, such as aerial topdressing, and the enthusiasm of the many returning young soldiers who were settled on farms after the war meant that production increased and we had more to sell.

Although Britain was not one of the original Common Market

members, it too introduced subsidies to encourage its own farmers to produce more. Synthetic fibres were developed by the industrial nations, and wool also faced competition in a market it had once dominated.

There was no single morning when New Zealand farmers woke up to find that their markets had vanished. The changed circumstances simply crept up on them gradually as production increased everywhere and competition became more intense.

A crucial time came in 1972, when Britain joined the European Economic Community. This was not a sudden decision, as it had been in the wind for a long time. One of New Zealand's most respected politicians, Jack Marshall (later Sir John), was assigned to negotiate a special deal for New Zealand in order to soften the blow. He, and others such as Agriculture Minister Brian Talboys, spent an enormous amount of time and effort in securing special arrangements. This was a particularly difficult task, as restriction of imports has been the dominant theme of the Common Agricultural Policy since its introduction.

Our negotiators did manage to secure some concessions, but only in conjunction with reductions in quotas on our exports of sheep meat and butter to Britain, and a total ban on cheese exports. Although our beef industry was much smaller then than it is now, beef exports to Britain did have some significance for us. These were also virtually shut out.

The initial effect of Britain's joining the EEC was masked by the quite buoyant world markets of that time, but the growth of the CAP and the Americans' counter to it increasingly impacted on New Zealand. We were told to diversify away from Europe, and the special access arrangements for butter and sheep meat were understood to be made to give us time to develop new markets. A central part of the deal was that Europe would not frustrate our efforts to develop markets in other countries.

This part of the deal was never kept, and a policy initially developed to make Europe self-sufficient in food was soon expanded greatly. As surpluses were created, export subsidies were introduced and the Community quickly became a large and aggressive food exporter.

New Zealand was continually frustrated by this policy. For example, after we were shut out of Europe, we developed a cheese market in Japan - but this market was promptly stolen by the Europeans, as they undercut the New Zealand prices and dumped their surplus cheese into Japan. Such unfair trade practices soon became the norm.

However unfair the competition became, we could do little about it. We had to accept that market conditions for New Zealand's traditional products had become much more difficult than they had been in the past. Unfortunately, the downward trend persisted.

There was a brief respite from 1972 to 1974, when a favourable exchange rate for New Zealand and a cyclical high point in United States beef prices coincided to improve the situation. The high beef prices brought a conflict within New Zealand, however, when the new Labour Prime Minister, Norman Kirk, proposed that export beef returns be used to subsidise the price for the consumer on the local market. The farmers protested vigorously; they won the day and the scheme was shelved. The government was the ultimate victor, however, as it then revalued the New Zealand dollar, and the impact on farmers' incomes was similar to that initially envisaged by government.

These internal wrangles serve only to show that while many of New Zealand farming's developing problems were caused by changing market conditions overseas, they were compounded by poor policy decisions at home.

What many people described as a dual economy emerged in New Zealand. There was one economy for those who depended on the international market for their incomes. These people took what the market provided, less the cost of getting produce to market.

There was a second economy for those who were supplying the home market with goods and services; they were protected from overseas competition and simply passed on the costs of production to the consumer. Wage earners and beneficiaries came into the protected side of the economy, as their returns were adjusted by the rate of inflation.

Not surprisingly, New Zealand's rate of inflation gathered momentum, and in 1972 we entered a period of 15 years when it was at least double that of our major trading partners. Inflation had a devastating effect on the farming community, but no-one was prepared to do anything about it for over ten years.

The National Party, led by tough-talking Rob Muldoon, was elected in 1975 with a mandate to take the hard decisions needed to right the economy, but in fact did very little with this mandate. Rob Muldoon appointed one of his ministers, George Gair, to cut expenditure soon after the election, but when these measures were strongly opposed by many (including farmers) nothing much more was done. Looking back, it is difficult to see why we let things get out of hand for so long.

In the 1960s the government had convened a National Development

Conference. One of its outcomes was the setting up of the Agricultural Production and Manufacturing Development Councils. The need to diversify markets and products was recognised, and various incentives were dreamed up to encourage this.

The main new incentives for both farmers and manufacturers that came from the development conference were tax-breaks. While these did persuade people to diversify, many of the manufactured exports generated were not market-related as they were sold at prices below their New Zealand cost of production. The New Zealand taxpayers and consumers paid the difference.

In the case of farming, low interest rate loans for development and subsidies on fertiliser and some chemicals were introduced. There was also some diversification. Beef numbers certainly rose as farmers were encouraged to use cull calves from the dairy farms to build up herd sizes. However, the main exhortation from government to farmers throughout the 1960s, seventies and even into the eighties was for increased production before diversification.

The assumption must have been that the overseas market downturn was a temporary aberration - that things would come right as they always had before. If we increased our production, we would be in a position to take advantage of the inevitable upturn when it finally arrived.

Farmers generally are cautious people. They do not willingly increase their borrowing or their workload if there is no return in it for them. The government found that incentives and subsidies had to be increased on farm inputs such as chemicals and fertilisers in order to persuade farmers to push up production every time we had balance of payments difficulties.

The reasoning behind the drive for more production, especially from sheep farming, was that countries with new-found wealth from the escalation in oil prices would queue up to buy our sheep meat. As synthetic fibres were oil-based, it was also thought that these would increase in cost and wool would have a new lease of life. Sheep therefore appeared a good bet, and many of the new subsidies offered were aimed at more sheep production (for example, the notorious 'skinny sheep' regime, referred to in the next chapter).

Settling young farmers on land and giving them development loans on favourable terms was also popular. Duncan MacIntyre, Minister of Agriculture in the National Government from 1975-84, was quoted as saying that one of the best tools for increasing production was a young farmer with a big mortgage. The implication was, of course, that this young farmer had to increase production to get ahead.

There were two main surges in this production drive. The first came in the late sixties. The farmers responded to the call, fertiliser application was high, and stock numbers increased dramatically. But they were quickly disillusioned because the prices for their products fell. By the end of the sixties many said that all they had seen for their efforts was more work and higher mortgages, with no additional income. A number considered that they had been duped, and said they would never again respond to calls for more production.

Amazingly enough, however, a few years later they did. Whether it was the better prices of the early seventies and for beef in 1979, the sheer size of the incentives provided, or a combination of both, farmers responded and there was another large surge in production in the late seventies and early eighties (see Appendix for Figures 1 to 3, showing fertiliser use 1968-95, stock numbers 1960-95, and product prices 1965-89).

While many of the incentives in the second production drive were aimed at the sheep industry, diversification was also encouraged and there was a spectacular rise in the amount of kiwifruit planted and in the number of deer and goats farmed.

At the end of all this, the reality was that most farmers were no better off than they had been before the production drives, and many considered themselves to have been actually disadvantaged by them. By using these policies we were creating some huge problems for the future, but no-one had yet realised this, nor how difficult life would become.

THE NEW ZEALAND SUBSIDY ERA

Subsidies crept up on the New Zealand farming industry as our international markets declined and inflation within the country escalated. They were gradually introduced in the early sixties, and steadily increased until 1984 when it was announced that most of them would be eliminated. By 1987 they had been phased out and the era was over.

Our farming and political leaders had argued for market access for our produce, declaring it to be unsubsidised, and even as subsidies were introduced they were considered to be very modest in comparison with those paid in other countries.

In 1982 the Guild of Agricultural Journalists held its international conference in New Zealand. We invited journalists from all around the world here to show them how efficient we were and why we deserved our place in the international market. But the exercise backfired on us, because once the journalists were here they assessed the true level of farming subsidies and found it to be particularly high.

In reporting back to their own countries many suggested that their governments need not give access to New Zealand products any more because the New Zealand Government was looking after its farmers. Farm leaders in the various countries fastened onto these reports and increased their lobbying against access for New Zealand.

When I attended the conference of the International Federation of Agriculture Producers for the first time in 1986 and announced that agricultural subsidies had been withdrawn in New Zealand, the French retorted angrily, 'Your leaders have always said that you were unsubsidised, so what is there to withdraw?'

Some government input into agriculture is universally accepted. Generic research, for example, is considered to be a necessary and important function of government.

The policy document adopted by the International Federation of Agriculture Producers' conference in May 1994 (*Farmers for a Sustainable Future*) states that government responsibility includes maintaining a commitment to support agricultural research. It notes that monies expended on research, training and extension fall into the 'green box' (allowable) category of support to agriculture under the GATT Uruguay agreement.

Ministry of Agriculture inspection services are paid by governments in many countries, and after much pressure they were paid, for a short period, by the New Zealand Government also. In our case, it was not so much that farmers considered that they should not pay for an efficient service, but that they were certainly not prepared to pay for the extravagant, unionised, uncompetitive system that had developed through the 1970s.

Ministry of Agriculture extension work is another service supplied free of charge in many countries, and this applied in New Zealand too. It was never seriously considered a subsidy. Governments saw it as an important link between researchers and farmers, and a necessary tool if farmers were to apply the research findings and increase production.

I doubt that many New Zealanders saw the settlement of returning soldiers on farms (on generous terms) as a subsidy. They saw it as a fair recompense to these brave young men who had devoted five or six of the best years of their lives to war service.

After the soldiers were settled, qualified young farmers were also financed into farms on interest rates which, while above the low returned soldiers' rate, were somewhat below market interest rates. Once again, this was not considered out of the ordinary. After all, many thousands of young New Zealanders were able to buy homes using government mortgage finance at low interest rates. Why should farmers be any different?

However, once begun, the subsidy system quickly proliferated as each sector wanted what the others had, and costs to farmers continued to rise. In 1976, following that year's Budget, the Ministry of Agriculture published the details of all assistance to farmers (*Assistance and Incentives for Farmers, 1976-77*). This document included a comprehensive range of loans, incentives, tax-breaks and cash payments, administered not only by the Ministry of Agriculture and Fisheries, but also by the Ministries or Departments of Education, Lands and Survey, Inland Revenue, Transport, Works and Development, Electricity, Forestry, Post Office, Rural Banking and Finance, and the Treasury.

Under the heading of 'Farm Purchase' were listed farm purchase finance at a reduced interest rate, special settlement loans for outstanding young farmers who could not raise the usual deposit, ballots for settlement and development of Crown land, grants to young people who had saved for farm ownership in special accounts, and suspensory loans to sharemilkers (written off if the recipient farmed the property for ten years).

Farm Finance and Development was the next category of assistance. Under this came the infamous Livestock Incentive Scheme, with a loan option of $12 per stock unit increased or a tax option of a deduction from taxable income of $24 per stock unit increased. The loan option was written off if the increase was maintained for two years. Also in this category were finance and tax-breaks for development, depreciation allowances on buildings, plants and machinery, stock loans, special plant loans, loans for development of marginal land, lucerne establishment grants, refinancing of farm debts, seasonal finance, nil livestock values for taxation purposes, and farm mortgage guarantees.

Fertiliser was a vital part of the rural economy. The assistance in this area included a fertiliser price subsidy of $25 per tonne, a spreading bounty, transport subsidies and loans for upgrading of airstrips, storage bins and access roads. There was also a tax-break scheme to defer the cost of fertiliser for up to four years from the year of expenditure.

Forestry, soil and water also featured strongly. There were forestry encouragement grants, tax deductions for woodlots, shelter belts or erosion control, and assistance for irrigation schemes, soil conservation, drainage, rural water supplies and prevention of pollution.

Noxious plant control was heavily subsidised, as were schemes for the eradication of animal diseases and the spraying of pipfruit with pesticides. Adverse events such as floods and droughts were covered by subsidies to regrass pastures, refund of freight costs for livestock transported elsewhere or for feed brought in, and loans and tax relief in cases of hardship.

A wide range of grants and other forms of assistance were also available in the education, transport, electricity, estate duty, agricultural contracting, farm training and transport areas.

Last, and perhaps most important, were the stabilisation schemes in the dairy, meat and wool sectors, which kept the prices paid to farmers from going too high or too low. These were in theory to be self-balancing, but in practice farmers were almost always in debt to them. When the schemes were discontinued in 1985, there was a large amount of money owing, some of which was eventually to be written off in 1987.

Right through the seventies, Federated Farmers pushed hard for government investment in the farming industry. A campaign called AGROW (short for 'agricultural growth') was set up, and those of us in leadership positions peddled it round the country, after getting advice from a public relations firm.

AGROW was the farming response to 'Think Big', which the National Government developed in the late seventies to counteract the increase in international oil prices. 'Think Big' involved expanding the steel mill and building a number of mostly energy-related manufacturing plants. It was a bonanza for the manufacturing and construction industries, and farming leaders intended to get their share of any government money that was being thrown around.

Sir Henry Plumb, the President of the National Farmers Union of England and Wales, was invited to New Zealand and Federated Farmers sought his advice on how to prepare for and conduct an annual negotiation with the government. He had great experience in this sort of thing in the UK and later in Europe. While it never became quite as formalised in New Zealand, the annual budget negotiations did take place in a similar fashion here. Basically the Federation was asking for compensation for cost increases.

One plan the Federation had previously presented was called the 'cost adjustment scheme'. The concept was that government should keep costs down, but if it did not then farmers should be compensated by an amount in line with the previous year's increases.

Everything appeared to be in the government's hands, and its payments were almost always based on increased production. The Livestock Incentive Scheme, for example, gave per head payments for increased livestock carried.

Livestock numbers went up dramatically. Some people claimed that many of the 'increased' numbers had been there all along, but farmers had not declared them in order to save tax. Even if there were some truth in this, there was a genuine large increase in stock numbers. The scheme became known as the 'skinny sheep scheme', as to get the subsidy payments, some farmers put on far more stock than their farms could run.

The ultimate sweetener was the Supplementary Minimum Price scheme, introduced in 1978. Farmers who had been through it all before and were reluctant to increase production again were finally seduced by these SMPs. The scheme guaranteed prices to farmers for several years ahead. I am sure that government saw SMPs as a very low minimum floor price on which they would never have to pay out. However, pay out they did, and it was SMPs that finally broke the bank. Although it was the Labour Government that ultimately dismantled the subsidy system, the National Government, in its 1983 Budget, had signalled the end of both SMPs and the 1 percent interest rate loans to support the stabilisation schemes.

The writing was definitely on the wall, and one of the most extraordinary periods in New Zealand's agricultural history was about to come to an abrupt end.

THE TROUBLE WITH SUBSIDIES AT HOME

'Farmer Uses Government Money to Buy Boat' screamed the headline. The article referred to a government suspensory loan of up to $1000 given to a number of farmers to compensate and encourage them during a difficult period in 1971-72. It was just one of many subsidies that became available. The money was interest-free, and repayable only if the farm was sold during the next ten years.

Most of this money was spent wisely on the farms, with payments being quite well policed by officials, but there was a lot of ill-feeling as some farmers, for various reasons, did not qualify for the payment whereas their neighbours did. The fact that one farmer bought a boat with the money and then boasted about it to a tabloid newspaper exacerbated the friction. One of the problems with subsidies — unfairness, or perceived unfairness — was already apparent.

In an attempt to avoid accusations of injustice, the government set up appeal committees, on which farmers were represented, to adjudicate on disputes over whether or not individuals qualified for the various subsidy schemes. Farmers tend to be very harsh judges of their peers, and while this involvement on the appeal committees meant that government certainly got value for money, it did absolutely nothing for relationships between farmers.

I served on several of these committees and found the work extremely difficult. In retrospect, I think it was the range of anomalies that arose, even at this early stage of subsidisation, that made me question the whole process later on. To take drought relief as an example, the few farmers who were farming large properties and who had the resources to withstand the dry period actually received most of the money, while a huge amount of time was spent in distributing small amounts of money to the remainder. These latter received too little to do any good, and the former far more than they needed.

Drought assistance usually took the form of subsidies on the transport of feed and livestock. The problem here was that the price of feed almost always went up soon after the subsidy was announced, and the proposed beneficiary was no better off in the long run. This is a simple example of how subsidies can distort the market. There are many more.

Making low-interest rate loans available to qualified people to buy or develop land would appear to be a good system, giving long-term benefit both to those involved and to the country. In practice, however, considerable evidence suggests that it has the opposite effect.

Our Reserve Bank now (1996) uses interest rates to keep down the rate of inflation. If prices go up, or show signs of increasing beyond the target guidelines, interest rates are increased, thus making it less attractive for people to borrow and buy. People wishing to sell must do so at prices that are economic to the borrower at the higher interest rate. The bank is not acting on hearsay. The farming experience has been that when interest rates went down, land prices went up, and vice versa.

Very few young farmers in New Zealand now support the concept of subsidising interest rates for farm purchase. They know that this puts up the price of land. In effect, therefore, the subsidised interest rates do not assist the young farmer who is entering the industry, but the older farmer who is selling out. This might be acceptable if the older farmer is reinvesting in agriculture. Generally, however, this does not happen. The money is usually invested outside agriculture where higher interest rates provide a better retirement income.

Subsidies designed to increase livestock and develop land had a real impact on land prices also, especially on land suitable for development. I can remember Allan Wright, President of Federated Farmers 1977-81, telling members how land prices in Canterbury rose following the announcement of the Land Development Encouragement Scheme. This of course meant that the subsidies were captured by those selling the land, not those who did the work of development.

Figure 4 (see Appendix) on land prices shows that there was a spectacular increase in land prices during the period of subsidisation.

The fact that subsidies were not applied evenly across the board also caused friction among New Zealand farmers. Sheep farmers received much more assistance than dairy farmers. The dairy industry made good use of the money made available through the Dairy Board for price stabilisation at an interest rate of 1 percent, but the sheep farmers received much more in development loans and on-farm support.

This was mainly because much of the assistance was aimed at developing marginal land, and dairy farming is not usually done on this land, whereas sheep farming is. Most industry observers would also consider that the dairy industry was better organised and that it marketed its produce more effectively. Dairy farmers therefore needed less assistance from government. There is a peculiar logic in a situation where the more you help yourself and the better organised you are the less you get from government, while the worse you perform the more you receive.

Subsidising one sector and not another, or even assisting one more than others, is bound to cause friction and resentment. This certainly happened in New Zealand. It is one of the strongest reasons why support for farm subsidies ran out, especially among farmers themselves.

A greater problem still is that very little of the money aimed at farmers actually stays with them. In the meat industry wages and costs rose alarmingly during the subsidy period and through the period when prices were guaranteed to farmers by government. Figures 5 and 6 (see Appendix) show that in this period sheep meat returns at farm gate (before the subsidy was added) dropped dramatically, even though the market price for the most part was actually rising. In a six-year period the processors' and marketers' share of lamb returns increased from approximately $1.50 per kg to almost $4 per kg.

The protection given to farmers' incomes took away the need and the will to fight for efficiency in marketing and processing; and, even though the meat processing industry had been delicensed in 1980, it took the removal of subsidies to drive the necessary change.

The 'picking of winners' by the government (eg, its huge support for sheep farming) turned out to be an extraordinarily expensive exercise for the meat industry, and ultimately for farmers. As farmers responded to the subsidies, sheep numbers built up from 50 million in the early sixties to a peak of 70 million in 1982-3. In the end there was simply no market for the meat and wool from this many sheep. Prices collapsed, and quality meat was actually rendered down for fertiliser in 1983 and 1984.

Expensive as this was, in the long term it was not the most costly part of the subsidy era. The meat companies spent many millions in expanding existing facilities and building new ones to cope with the increased demand for killing space.

Many of these facilities then became redundant when sheep numbers fell again. In subsequent years, two major companies and several smaller ones failed, owing millions of dollars to creditors, including farmers. An elaborate bail-out scheme was devised for another major company, but this was in reality a failure as it left the buying farmer companies carrying too great a debt load.

Workers in both meat processing and inspection, whose wages and conditions had been established at unrealistic levels under farm subsidisation, were also forced to make painful adjustments when the reality of the marketplace had to be faced. It has already taken the meat industry ten years to adjust to the distortions caused by subsidies and protection, and their subsequent withdrawal. The adjustment may not yet be over.

The same sort of adjustment took place in the aerial topdressing industry, which geared itself up with expensive aeroplanes and support machinery to cope with a massive subsidy-inspired increase in demand. When the subsidies were withdrawn, it was then left with very little work and huge debts.

The subsidies for aerial spreading were high enough to encourage its use on flat and easy rolling country, when tractors and trucks should and could have been used. This situation caused considerable tension between ground and aerial spreaders of fertilisers. It is just another example, repeated over and over in other farm service industries, of the way in which subsidies distorted the market and caused tensions in sector relationships.

Town/country relationships also suffered throughout the subsidy era. In many respects the farmers' problems were caused by the protection of the manufacturing industry through licences and tariffs, compulsory unionism and national wage setting, and total protection of the public sector. This protection increased farmers' costs and made them envy their city friends.

However, the paying of subsidies to farmers turned city people against the country folk to a much greater extent. Despite their cash-flow problems, some rural people were still asset-rich by city standards. Many town people saw farmers as 'fat cats' who were becoming even wealthier at the expense of the struggling city taxpayer.

As a Federated Farmers leader, I spent much of my time defending and promoting what we called 'investment in farming'. We quoted figures showing how much in overseas funds and how much employment our industry provided. We spent hours writing articles and appearing on radio and television.

What we were saying was true enough. Eighty percent of the country's export earnings came from the land. It was probably even valid to argue that the production base should not be undermined through lack of investment. The industry was also a very large employer — 30,000 people worked in the meat industry alone. These were of course the government's prime reasons for subsidising farming — it was a good way to lift the whole economy.

Despite the best efforts of both farming leaders and government MPs, the public relations campaign never quite succeeded. For farmers, the government payments were never enough to compensate them for double-digit inflation, or for the cost of protecting other sectors of the economy. On the other hand, city people continued to regard farm subsidies as unfair and undeserved.

Fuel used on farms was exempt from road tax, and farmers were often accused of using it in their cars. There was actually talk of colouring the farm fuel so that offenders could be caught, and our city friends joked that they could 'smell' the subsidised petrol in our car exhaust fumes. We were also accused of counting sheep twice when per head payments were made.

While there may have been a grain of truth in these accusations, most farmers were honest and bitterly resented them. The town/country gap widened, and became one of the worst legacies of the subsidy era.

THE TROUBLE WITH SUBSIDIES ELSEWHERE

I have often heard it said that New Zealand is campaigning for the removal of subsidies worldwide simply because we cannot afford them ourselves, and we want everyone else to be on the same footing as we are. Someone using this argument usually goes on to say that his or her country can afford to pay subsidies, and that New Zealand should not try to impose its theories on others. A British journalist, Robert Forster, took this view at the World Sheep Conference, held in England in July 1995, when he said:

"If there were no hill farmers, then the hills would be empty and the taxpayers of this country have made it clear that they do not want them to become echoing wastelands.

They want people and houses and roads and services when they temporarily quit their cities for a breath of rural life. They may not know much about sheep, but they do know that the only people who can stick life in the hills are those who are bred to it.

We are a rich western country and we can afford to encourage these survivors to stay. We need them to keep the cobwebs out of our most beautiful areas, and much more importantly in my opinion, the national balance sheet shows we need their sheep.

Those countries which have dropped subsidies did not suddenly volunteer to become shining white free traders - they were compelled to because they have a small industrial base and simply cannot afford them.

And I think they are wrong to force arguments that are pertinent to their systems on faraway countries. We are all for freer trade - that is the way the world is moving - but the only reason they are beating the no-subsidies drum so loudly is because they know that if they stand alone they are at a huge disadvantage."

Forster makes an enormous assumption when he suggests that the British taxpayers support the payment of subsidies. I am not aware that they have ever been asked for their opinion. They have certainly not had the costs of the subsidy policy spelt out to them, in comparison with the alternatives. I am also unaware of any referendum held in Europe on whether or not the Common Agricultural Policy should continue.

New Zealand does not wish to impose its theories on others in order to gain an advantage. We do, however, want others to understand how their subsidy systems adversely and unfairly affect us and countries like us. When this happens, the 'domestic affairs' of other countries become our business, and we are certainly entitled to state our case.

Countries that pay farm subsidies can, to a large extent, afford to do so because of the money they earn from industrial exports. Barriers to exporting these products have been progressively reduced over the last 20 years, to the point where the goods are now traded quite freely.

It is, in my opinion, totally unacceptable for industrialised countries to argue for free trade in industrial goods while they continue to restrict imports of agricultural products. Forster's somewhat muddled argument (supporting both freer trade and subsidies) merely perpetuates this double standard. The best reason he can come up with to justify his stance is that he lives in a rich western country - not exactly a logical argument, and certainly not a politically correct one.

It is also untenable for industrialised countries to export surplus agricultural produce with the assistance of export subsidies, to compete against produce from more efficient and unsubsidised farming nations. It was once thought that such surplus produce could also be used to feed hungry people, but this idea has now been discredited.

These issues will be dealt with much more fully in subsequent chapters.

The cost of subsidising agricultural products in industrialised countries is extremely high. Much of the cost is ultimately borne by other agricultural exporting countries, but there is also a very high cost to consumers and taxpayers in the countries where subsidies are paid.

There are three main types of agricultural subsidy: production subsidies, border protection, and export assistance. Production subsidies can support both inputs and outputs in agriculture and include subsidies on chemicals and fertilisers, low-interest rate loans for development or price stabilisation, and government support of product prices. In the main they are a simple transfer of funds from taxpayers to farmers.

Border protection is used to keep out competing products from other countries, usually by the imposition of high tariffs. It forces consumers to buy locally produced food, and is a transfer of funds from the consumer to the farming sector. The most extreme examples of border protection policies are practised in Norway, Sweden, Finland, Switzerland and Austria. The Japanese rice industry and the Canadian dairy industry also fall into this category. These countries reason that although food production is very costly, they still wish to produce their own food and therefore must keep out competition. The jargon term for such activity is 'food security'.

Export assistance follows on from the other two forms of subsidy, which almost inevitably lead to surplus production. This surplus is mostly disposed of on the world market, where prices are lower than the protected internal ones. Someone has to fill the gap between what farmers have been paid and what the international market provides. It is usually the taxpayer.

Government money is also spent on storing surpluses through periods of low prices. The wait for higher prices can be a long one, and the storage cost enormous.

The CAP costs the consumer a huge amount. In the *Daily Telegraph*, 20 October 1995, British Labour MP Stephen Byers commented on figures obtained from the Organisation for Economic Co-operation and Development (OECD), and gave some examples. He said that a pint of milk currently cost 36p at farm gate, but without CAP tax it would cost only 17p. Consumers were also paying more than 50 percent CAP tax on some of the most popular types of beef. Sugar, which cost 71p per kilo, would cost only 33p without the CAP. Other foods which were much more expensive because of the CAP included bread, flour, chicken, pork and lamb.

The CAP, which accounts for the lion's share of the whole EU budget, already costs £40 billion per year, despite attempts to reform it. Two aspects of the CAP which enrage Euro-sceptics are huge amounts paid to European tobacco growers - most of whose product is so substandard it has to be sold outside the EU - and to makers of substandard wine, most of which is turned into industrial alcohol.

The National Consumer Council in the UK estimates that the cost of the CAP in consumer terms is £10 per week per average family. This of course does not include the cost per family in extra tax paid. There are also many hidden costs. These come from fraud, pollution and wastage, and the cost of the bureaucracy required to support and police the systems.

The size of the bureaucracy needed to police the CAP, the US Farm Programmes, and other protected systems is massive. As the systems have become more complicated, now including quotas on production and set-aside of land, the red tape required has multiplied every year.

Fields have to be measured, livestock counted and milk weighed. Farmers in the UK tell me that they regularly have to fill in up to 80 pages of statistics forms. They appear to spend almost as much time farming the subsidies as they do farming the land.

Despite the existence of this army of bureaucrats, fraud is still a problem. Anthony Rosen, a British journalist and consultant, told the *Southland Times* in November 1993 that a major downside of the CAP system had been the rise of fraud, not so much by farmers as by bureaucrats, marketers and processors. Fraud had at that time been estimated to cost £6 billion per year, and Rosen thought this was on the low side. A more recent assessment puts it at £10 billion annually.

Wastage is also a costly problem. The issue of the *Daily Telegraph* referred to above also disclosed that in August 1995 Eurocrats had ordered the destruction of more than two and a half million tons of fresh fruit and vegetables in order to keep shop prices artificially high. Brussels paid farmers £439 million, £57 million of which was contributed by British taxpayers, for the pulped fruit and vegetables, most of which came from France and southern Europe.

There is of course some wastage in a free enterprise farming system. Unforeseen circumstances such as unusual weather can lead to gluts of produce. The cost is much lower in a free situation, however, as the loss will usually be taken before the surplus is harvested, processed and stored.

Pollution has been a problem in many of the highly subsidised countries. It is not totally a product of subsidies, but the high-priced regimes have almost always led to high input/high output farming and subsequent pollution of waterways.

There is no doubt that subsidies are a considerable cost to consumers and taxpayers in the country concerned, and to farmers in other countries. They have almost certainly increased waste and pollution, and they have made farmers and consumers subject to the domination of bureaucrats and politicians.

The farm service sector may have been better off, but it is debatable whether most farmers are. The support, no matter how generous, has not always ended in a better bottom line.

I remember being with Gerry Thompson, then New Zealand Ambassador to Belgium, at Leuven University, just outside Brussels, in 1990. We were debating the merits of farming without subsidies with about 40 senior agricultural students and their lecturers. Gerry and I were somewhat outnumbered, but thought we were giving a good account of ourselves, when a guest from one of the Belgian farming organisations asked if he could join in the debate.

Gerry and I thought that he would weigh in with a strong speech in favour of subsidies, and braced ourselves for the onslaught. We were completely surprised when he took our side of the argument, pointing out to the audience that the most profitable sectors of Belgian farming were those not dominated by EEC regimes. He was referring to flower growing and pork and poultry farming.

It is notable that the unsubsidised sectors worldwide seem to have achieved greater technological and efficiency gains than those that are subsidised. This may be coincidental, but I doubt it.

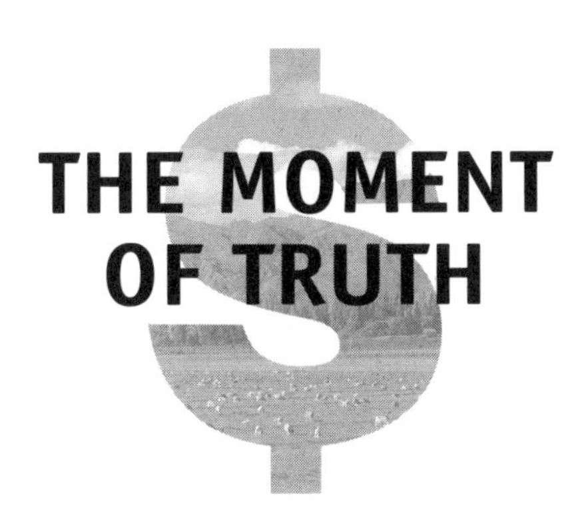

THE MOMENT OF TRUTH

The moment of truth for Federated Farmers leaders came in 1982, well before it came for most other people, and two years before the politicians began to take action.

For some years the Federated Farmers Budget submissions were referred to by Duncan MacIntyre (Minister of Agriculture, 1975-84) as 'the farmers' shopping list'. While this was a standing joke, in reality there was very little difference between the National Party manifesto and the 'shopping list' in any given year. Most of the negotiations concerned the actual carrying out of the manifesto. As Duncan MacIntyre often put it, the government was 'having a bit of trouble getting the funds from Treasury', and not everything could be done at that particular time.

In 1982, however, the government got a very different 'shopping list' from the Federation. Late in 1981, Federated Farmers President Rob Storey had asked me, as junior vice-president, to convene a committee to prepare the submission for the 1982 Budget. My colleagues on this committee were the vice-chairmen of the three produce sections: arable, dairy and meat and wool (David Ritchie, Alan Bailey and Bruce Anderson respectively).

We started by measuring the actual cost increases farmers had incurred, with the intention of seeking compensation. The country was going through a period of particularly high inflation, and we soon discovered that the amount needed to compensate farmers fully was enormous. We knew that the country could never afford to pay the compensation, no matter how well we could justify it.

We also realised that deficit budgeting by the government was the greatest single cause of the inflation: compensation to farmers would be likely to increase the deficit and consequently increase the rate of inflation. The whole process was a vicious circle in which any gains we made would be quickly eaten up again by inflation.

The committee decided that the time had come to attack the cause of the problem rather than to seek compensation for it. Simply seeking subsidy increases for one more year would mean a greater problem the following year, and the whole thing would snowball. We reported back to Rob Storey and the executive committee our view that a completely different approach was needed. With their backing, a submission was produced, which showed that:

- Overseas earnings had increased from $3,034.3 million in 1979 to $5,325 million in 1982.
- Stock units had increased from 99 million in 1979 to 109.5 million in 1982.
- At the same time, terms of exchange at the farm gate for both dairy and sheep farmers had steadily deteriorated.
- Farm costs had increased by over 80 percent from 1979 to 1982, but returns had increased by only 56 percent (including subsidies) in the same period.

The following statement was emphasised in bold type:

The key to continued investment in agriculture is profitability, and the major impediment is the current high level of inflation. In terms of agricultural policies, the Federation submits that if inflation were controlled at a level at least comparable to that of our major trading partners, the requirement for government assistance to agriculture would be negated. The Federation considers that the reduction in the level of inflation is the major policy objective for 1982, and would, if government adopted a planned package to achieve this end, accept the same disciplines as would be required by other sectors of the community.

Our full economic package was presented to the government in February of 1982, and was summarised by Rob Storey in a press release on February 22, in which he stated that the 1982 Budget should attack causes of inflation rather than symptoms. He cited the underlying causes of New Zealand's high inflation rate as inadequate wage-fixing policies, high government spending, protection for many sectors of the economy, and licensing regulations that limited competition.

He said that if the export sector was to survive and produce overseas funds, decisions had to be made immediately to ensure the sector's competitiveness on the international market. If decisions to correct the economy were delayed, a slide - probably worse than anything previously experienced - could start and would be hard to stop. Certainly the current growth rate would be checked, and productivity would fall in all sectors of the community.

A main cause of inflation, government expenditure, which had reached more than 39 percent of GDP, was an obvious area for urgent correction. Statistics showed that total government expenditure had increased by 2 percent in real terms in 1980-81, with a similar increase projected for 1981-82. This necessitated increased borrowing and tax revenue. The Federation fully supported the user-pays principle and a maximum private enterprise economy.

As a contribution to holding government expenditure, the Federation would not seek an increase in SMPs for the following season, and expected other sectors to show similar restraint.

The Federation did believe, however, that increased development funds were required to maintain growth in agriculture, and suggested extra loan schemes for specific areas such as electric fencing and water reticulation, in order to increase production. A continuation of fertiliser and transport subsidies was strongly recommended.

While these submissions were a major departure from the approach taken in previous years, they were still somewhat of an 'each-way bet'. There was a concentration on pushing down the rate of inflation through reduced government spending and introducing more competition into the economy. The offer to forego any increase in SMPs was the farmers' contribution to what had to be an across the board attack on inflation. The farmers stopped well short, however, of advocating the elimination of subsidies, as can be seen by the requests for new development schemes and continuation of fertiliser and transport subsidies.

Prime Minister Rob Muldoon's response to the submissions was interesting, to say the least. He gave us a lecture on economics, Muldoon-style, told us not to bring our 'silly ideas' to him, and advised us to 'take the subsidies and run'. The gist of his lecture was that the government taxed people who had more money than they needed and paid it to those who deserved to have it. Farmers deserved some of it because they provided the overseas funds that the country so badly needed.

While those of us who presented the submission were shown the door by the Prime Minister, the government did respond indirectly to the call to control inflation with the introduction of a wage/price freeze.

This was a very different approach from using competition to bring costs under control, which the farmers advocated. However, the Federation supported the freeze as an interim measure to bring home to New Zealanders just how important it was to control inflation. We thought that the freeze might be a useful stopgap until more fundamental measures were taken.

The government, however, did very little other than implement the freeze ever more strictly over the next two years. The nominal rate of inflation fell quite dramatically, but the basic problems of deficit budgeting and protection through licensing (with the exception of transport) were not dealt with at all.

The only changes the 1983 Budget signalled were the end of the SMP scheme and the availability of 1 percent interest rate loan money to the producer boards, and the transport deregulation.

It was quite different in Federated Farmers. Our organisation was suddenly seen in a new light by economic advisers from both public and private sectors, a number of whom offered their assistance. Over a two-year period a new policy was developed, after extensive consultation with these experts and our own membership.

Once we had our policy, we planned to market it to the political parties, with the objective of having it incorporated into each party's 1984 election manifesto. *Agriculture - the Anchor of the Economy* was born. Many consider this was the most important document ever produced by Federated Farmers. It covered a wide range of policies on agriculture and the New Zealand economy, the most important relating to the domestic economy, protectionism, taxation, transport and industrial relations.

For the domestic economy, the Federation sought:

- policies that would attack the causes of inflation and ensure that its level is reduced and consistently maintained below that of our trading partners
- the establishment of a realistic exchange rate, adjusted to reflect the value of the NZ dollar
- the progressive removal of all protectionist devices
- minimum government intervention in the capital market
- a wages policy based on free bargaining, with wage movements based on productivity and the ability of the industry and nation to pay
- tax reductions in lieu of wage increases
- reduction in public sector debt; savings made in the overheads and administration of government departments; the opening to public shareholding of government agencies acting commercially
- resources to develop export efficient industries
- competition legislation in the areas of mergers, takeovers, monopolies and restrictive practices

On the subject of protectionism, the Federation asked for urgent implementation of government policy to reduce tariffs progressively, to review regulations affecting commercial freedom, and to push for the reduction of world protectionist measures.

The Federation also asked for a real reduction in total tax collection by the government, a flatter rate of income tax, a system of tax averaging, and regular updating of estate duty exemption.

In the transport area, it sought progress towards free competition between road, rail and air; rationalisation of resources involved in the shipping industry on-shore; freer competition among shippers for New Zealand cargo; and the use of road tax funds for roading.

The Federation sought the adoption of an industrial relations policy which provided optimum job opportunities; obtained maximum efficiency and productivity; provided remuneration based on skills and the ability of country and industry to pay; allowed employers to hire whom they chose and to terminate employment subject to the Industrial Relations Act, and to operate free wage bargaining within a system of conciliation and arbitration.

On the Supplementary Minimum Prices issue, the Federation said: 'The continuing need for SMPs would be eliminated if the economic package advocated by the Federation was adopted by Government.' This was the major point of the whole policy - get the basics right and subsidies will not be necessary.

It was also made clear by Peter Elworthy, Acting President of Federated Farmers when the document was launched, that the policy was a total package - not something that politicians could pick and choose from. Picking and choosing was, of course, exactly what did happen, and the consequences of this have been debated ever since.

Time did not allow *Anchor of the Economy* to be incorporated into any political manifesto. The 1984 election was expected in November, but the Prime Minister called a snap election for July. The only group with anything resembling an economic manifesto was Federated Farmers, but it was not contesting the election.

The Labour Party won the election in a landslide, and a new era in New Zealand politics and commerce began.

THE PROCESS OF REFORM

A number of points need to be made about the process of reform in New Zealand. The first is that not only agriculture was involved. The second is that everything did not happen over night, as many people seem to think it did. Reform has taken more than ten years so far, and will probably continue for some time yet. The third point is that despite assertions to the contrary made by Roger Douglas, farmers did not have to be dragged kicking and screaming into the process. Farmers were advocating reform before any New Zealand government became involved.

The reform period that commenced with the new Labour Government's 1984 Budget was difficult for the farming leadership. There were very vigorous arguments between Federated Farmers and government members. There were also arguments among farmer organisations, bitter disputes between farmers, trade unionists and manufacturers, and robust debates between the Federation and the National Party.

Federated Farmers is a non-party political organisation but over the years farming voters have mainly supported the National Party. Some people therefore saw the Federation as an extension of the National Party, especially as many people were active in both organisations. I have many times heard the Federation referred to as the 'National Party in gumboots', despite the fact that it had officeholders and supporters who were not members of any political party, and others who came from across the political spectrum.

A particular difficulty for the leadership of the Federation was that many farmers considered the Labour Party to be the sworn enemy, even if it was doing the right thing. Many National Party members were in a dilemma because while their party was in opposition they believed that it was their duty to oppose most of the reforms - even if they privately agreed with them. We often ended up with Federated Farmers supporting policies that were opposed by the National Party, and Peter Elworthy and I were given a very hard time for 'supporting the Labour Government', when what we were actually doing was supporting policy and reform.

Our arguments with the Labour Government of the time were rarely about policy or the reforms themselves, but about the fact that only some of the necessary package had been carried out. About this time, Owen Jennings of Federated Farmers and Trevor de Cleene of the Labour Government had an acrimonious debate at a meeting in Invercargill. De Cleene stated that his government had carried out the Federation's policy. Jennings pointed out that the policy was a total package, and implementing part of it was not good enough. De Cleene's response was to walk out of the meeting.

The major announcements concerning the elimination of farm subsidies were made in the Labour Government's first Budget of 1984: fertiliser and noxious weed subsidies were to be phased out, irrigation and water supply subsidies reduced, cost recovery for product inspection services reintroduced, and Rural Bank interest rates increased progressively until they reached market rates.

Several taxation changes were also made, and it was announced that a tax on spending - a goods and services tax (GST) - would be introduced in 1986. The tax was introduced at 10 percent and increased to 12 1/2 percent in 1989.

The initial impact of the changes made in 1984 was not particularly dramatic for farmers. Costs were still influenced by the price freeze, while the 20 percent currency devaluation made at change of government improved returns at farm gate, offsetting the removal of SMPs. Many farmers were financed by the Rural Bank, and the initial increase in interest rates could still be managed under the prevailing conditions.

Some have described 1984 as the calm before the storm, but it was actually a time of total euphoria for many New Zealanders. A large portion of the business sector considered that the new government was a breath of fresh air in that it was prepared to sweep away the unpopular controls imposed by the previous government. The Bob Jones-led 'free enterprise' New Zealand Party had taken support away from the National Party and helped Labour to win the election.

The trade unions too saw 'their' party in power for the first time in nine years, and looked forward to the end of the wage freeze and to reductions in unemployment.

The Prime Minister chaired an Economic Summit Conference at which people from all sectors of the economy were invited to put forward their ideas for restructuring. I was pleased and proud that I had been able to get the statement that 'export-led growth' was what New Zealand required included in the conference communique.

Several other 'summit' conferences were held, including one on employment. In the end, however, they turned out to be no more than public relations exercises that had little relevance in subsequent years.

Federated Farmers, for example, had many arguments with government advisers who claimed that export earnings were irrelevant, despite the fact that export-led growth was an agreed outcome of the Economic Summit Conference. If I heard the statement 'capital inflow can replace export earnings' once, I heard it a hundred times.

Once controls were lifted, it did not take long for all the undesirable features of the economy, which had been held at bay by the wage/price freeze, to rear their ugly heads again. Unions, naturally enough, aimed at as much as they could get in wage increases, and those who had had prices frozen also looked to make up for lost ground.

Costs began to increase spectacularly. Despite the government's goal of medium-term growth through price stability, inflation reached 16.6 percent per annum in June 1985. It fell slightly for a short period, but rose again to 18.2 percent in September 1986, following the introduction of GST. It was to be December 1987 before the inflation rate fell below 10 percent again.

The New Zealand dollar was floated on 4 March 1985. I was a supporter of this move, as I felt that the market would continue to devalue our currency if inflation persisted, and that floating the dollar would enable exporters to hold their own. But this did not happen. The government operated a policy of tight monetary control, with associated high interest rates. These rates attracted money into the country, and the exchange rate went up, rather than down.

The high interest and exchange rates really affected only those in the export sector. With the controls off prices and wages and our secondary industry still protected, most other sectors could pass cost increases on.

Many things which were to have a positive effect in the medium term, such as the creation of efficiently run State Owned Enterprises, were being put into place. These were set up to replace inefficient government departments, but it would be some time before they would make a positive impact.

Figure 3, (see Appendix) which shows terms of exchange indices for the pastoral industries, gives a good indication of the decline in farmers' fortunes in this period. It shows very clearly the drop between 1985 and 1986, when the effect of the initial changes began to bite. Indeed, 1986 was a year of real anger in the farming community.

Farmers, especially those not financed by the Rural Bank, faced huge increases in their interest rates. Rates of well over 20 percent were not uncommon, particularly for seasonal finance. Farmers who had borrowed heavily in the production drives of the previous government were hit extremely hard.

In addition to the removal of support announced in the 1984 Budget, tax changes were made which took away concessions from farmers and others. Prices of some land and livestock had fallen, and a number of farmers were technically insolvent.

Prime Minister David Lange predicted that up to 8,000 farmers could lose their farms in the restructuring process. Farmers were understandably angry and there were a number of demonstrations including one widely reported occasion when the Prime Minister was at a function at a research station near Dunedin. Owen Jennings was at a farmers' meeting nearby, and was asked to try to arrange a meeting with the PM for a delegation from the farmers' group. The Prime Minister refused to meet the delegation, upsetting some of the farmers to the point where they took out their frustrations on the PM's car, which was slightly damaged.

Some financial organisations moved to foreclose on farmers who were in arrears to them. Irate farmers mounted pickets at mortgagee-sale auctions, in an effort to stop what they felt to be unfair sales.

Much negotiation took place between farmers and the government on this issue, and in July 1986 a mortgage discount scheme was announced through the Rural Bank, in an attempt to keep viable as many farms as possible.

During the period of the Meat Industry Stabilisation Scheme, farmers had built up a debt of $1 billion. This debt was ultimately written off by government in March 1987, after a very difficult negotiation. One problem was that the Meat Board had money in another account (the Meat Industry Reserve Account) which came from the British Government in recognition of New Zealand's supplying of meat to Britain during World War II. The government wanted this money in return for writing off the other debt.

The Meat Industry Reserve Account had been used by the Meat Board as a loan account to assist in the establishment of co-operative fertiliser and meat processing companies, and farmers were not happy at the prospect of losing it. They claimed that the two funds were totally unrelated, and that the stabilisation debt was as large as it was only because the previous government had set minimum prices unrealistically high.

In the end a compromise was reached. The debt was written off in return for part of the Meat Industry Reserve Account. While this write-off amounted to $40,000 per farmer, it did nothing to help the farmer's plight at the time except to remove the worry of the debt being called in later on.

During the first three-year term of the Labour Government, which had been so difficult for farmers, most other people in New Zealand did very well financially. Some had a right royal time, and the day of the 'yuppie' was at its height. It may have been an over-reaction to the pent-up demands of the freeze but the property, share and financial markets all boomed after the controls were removed, and many fortunes were made. People in the workforce had substantial wage rises, and manufacturers were able to increase their prices also.

Farmers, who were working very hard either to stand still or to slow the slide backwards, resented the whole process. There were marches in the streets of most rural towns, culminating in a major demonstration in Wellington in 1986. These demonstrations were not in favour of a return to the old systems, however. Farmers knew that there had to be change. They simply felt that so far the reforms had discriminated against them, and them alone.

In his final address as Federated Farmers President in 1987, Peter Elworthy, frustrated by the government's apparent inability to see this injustice, told the conference that the farmers had been 'taken for simple suckers'. Peter has told me since that he took this line from a speech by an earlier Federated Farmers President, Bill Dunlop. It was obviously not a new phenomenon.

Not surprisingly, the Labour Government was returned with an increased majority in the 1987 election. Many people were better off, and those who were not considered that Labour needed to finish the job it had started.

The world share market collapsed just before the election. Our share

A number of cartoons about Prime Minister David Lange and his relationship with farmers appeared during the reforms. The dog-tucker cartoon from Straight Furrow caused considerable controversy at the time. Cartoons courtesy of Glenys Christian.

market was badly affected also, but the collapse came too late to impact on the election.

It has been said that the New Zealand Labour Government has been the only government of recent times ever to be re-elected after introducing a tax on spending. In many respects the re-elected government's real work was about to begin. The share market collapse had far-reaching effects, and many of the newly made fortunes were lost again.

As far as farmers were concerned, the fact that people in other sectors were finally being forced to face up to the real world dissipated a lot of their anger. With the collapse of the share market, we were back in the real world. The notion that capital inflows could replace export earnings also disappeared and has seldom been referred to since.

Export-led growth became the goal again, and the export sector began to be valued once more.

STRATEGY FOR THE FUTURE

Strategy for the Future was the document Federated Farmers produced for the 1987 election. Its opening paragraphs, entitled 'Meeting the Challenge of Change', summed up the farming community's mood at the time and its goals for the future:

"New Zealand is caught in the turmoil of a changing world. Federated Farmers believes that there is no turning back for this country. The only way to go now is forward. That may not always be easy, and it presents a considerable challenge. Agriculture has accepted that challenge, and is determined to meet it with vigour and dynamic innovation.

The future living standards of all New Zealanders will depend on export-led growth. Agriculture has greater potential than any other industry to provide the country with this growth.

For more than a century, land-based industries have capitalised on the natural advantages of climate and soil, enhancing these assets by skilful use of science and technology. This innovation and enterprise has meant that the agricultural industry has been able to overcome the handicap of distance from major world markets.

Family farming has played a tremendous role in this development, with women making a great contribution to the industry.

But in the last decade soaring inflation has eroded all exporters' ability to compete on the world market. The economy will not prosper until inflation is below the level of that of New Zealand's trading partners."

The document asked for the adoption of commercial principles in running government departments or corporations; the elimination of protectionist devices such as tariffs, import licences, and export incentives; and the pursuit of international trade reform. It recommended repealing provisions that supported trade union monopoly, basing wage increases on productivity, and creating equitable taxation schemes.

The Federation went on to detail policies on a floating exchange rate, accident compensation, the family farm, the Rural Bank, producer boards, research and development, live animal export, the ports, and local government. It concluded with another strong statement on the economy:

"The future prosperity of agriculture is tightly linked to sound economic management, which means continuing the process of change. The strategy for growth should embrace change in social and economic environments and in attitudes. A competitive, market-related economy must be created by eliminating inflation through a low internal deficit, achieved by reducing government spending, and dramatically reduced protection and a free labour market.

The freehold family farm will continue to be the cornerstone of the farming industry, and will act as a springboard for future change.

Federated Farmers is prepared to meet the challenge these changes offer in order to build a dynamic nation for the future."

In retrospect I think it is fair to say that this was a farsighted policy statement, and most of its recommendations were carried out over the following five years.

It was much firmer on the removal of subsidies than *Anchor of the Economy*. It still argued for a concessional interest rate for young farmers, and for government funding for basic research. However, these were the only areas where any form of government subsidy was suggested, except perhaps for some recommended tax deductions. Here a write-off was proposed to assist farmers to move to the new system of taxing livestock. As with the young farmer concession, this was only a transitional measure.

The advocacy of a freely floating exchange rate rather than the managed float recommended in the previous document was interesting. The Federation believed that this would impose a real discipline on politicians. As the paper itself said, this 'would only deliver benefits to the economy if the fundamental economic reforms were carried out'. The whole concentration of the document was on getting economic management right.

Over the years, the New Zealand Government had built up a huge portfolio of businesses. It owned transport businesses (including railways, road services, an airline and a shipping corporation), electricity generation facilities, postal and telephone services, hotels and travel agencies, a printing and publishing business, television and radio stations, a steel mill, a fertiliser company, finance and insurance companies and banks. These were in addition to services such as health, education and social welfare, which are run by the governments of most countries.

The staff of most of these companies were public servants who worked in a very protected environment. It was virtually impossible to sack a public servant. Company profits were low, and some companies had large losses every year. Government ownership of many of these businesses was a real drain on taxpayers' funds, and they were one of the causes of the high inflation rate.

The Labour Government, in its second term of office (1987-90), made great progress in this area. Most of these organisations were corporatised, and a number sold to the private sector. This brought about huge improvements in efficiency. Railways, for example, as a government department had over 20,000 people on the payroll. After corporatisation the staff was reduced to 5,000, but the slimmed-down organisation still managed to move more freight than ever before.

Money obtained from selling state businesses was used to reduce debt. Government departments that were not sold had their budgets reduced, and they too were forced to become more efficient. The public service stranglehold on the economy was finally broken.

A four-stage programme of tariff reduction was introduced, and a two-step cut in motor vehicle duty announced. It could be argued that the tariff cuts were too little and too late, but at least the process was begun, despite loud protest from some manufacturers and union representatives.

Costs incurred on the New Zealand waterfront had long been a major cause of irritation. Watersiders worked approximately half the hours of other workers, for twice the pay. This was a national outrage, and a courageous Minister, Bill Jefferies, took up the challenge and reformed the ports. Federated Farmers was fortunate in having Malcolm Lumsden as the hard-hitting coordinator of its efforts to lead the charge for this reform.

These reforms were all necessary, but they did of course lead to a drop in support for the government as the economy slowed down and as people who had traditionally been Labour supporters were affected personally. The rise in unemployment figures at that time was of concern to many people.

Rifts developed in Cabinet. Roger Douglas and Richard Prebble, two of the major reformers, lost their Cabinet posts, and David Lange resigned as Prime Minister when Douglas was re-elected to the Cabinet by the caucus. The main argument appeared to be about the pace of reform. Douglas was keen to proceed quickly, whereas Lange wanted a period of reflection, which he described famously as 'having a cup of tea'.

At last the policies began to work. By March 1989 the inflation rate had fallen to 4 percent. A Reserve Bank bill was introduced that year which set the achievement and maintenance of price stability as the bank's major objective. The Budget of 1989 set the target inflation rate at 0-2 percent by 1992.

Much progress had been made, but the job was not quite complete. Government expenditure was still very high, especially in the area of social welfare, and the Budget showed a deficit before asset sales. Labour market and tariff reform was also incomplete.

The Labour Government had lost much of its support, but the National Party still felt that it needed to make substantial promises in order to win the 1990 election. The most unrealistic of these promises was that it would remove the superannuation tax surcharge, which Labour had introduced. This surcharge was an unpopular but necessary measure to take in a time of large budget deficits.

National won the 1990 election in a landslide, but soon found that it had a crisis on its hands with the virtual insolvency of the still state-owned Bank of New Zealand. The new Minister of Finance, Ruth Richardson, took the elimination of the budget deficit very seriously and was able to achieve this by pruning social welfare spending. Needless to say, the promise to remove the tax surcharge on superannuation was not kept. The reason given was that the Bank of New Zealand crisis meant that the money was no longer available.

The labour market was at last freed up, with sweeping reforms taken in the Employment Contracts Act. The Accident Compensation Scheme was reformed to the point where it became more affordable for employers, who actually pay for it. Privatisation of the state-owned enterprises continued apace.

The result gave New Zealand an inflation rate consistently less than 2 percent and made it one of the most competitive countries in the OECD. In November 1995, it had the third lowest rate of unemployment in the OECD. The last few years have produced government budget surpluses, and plans are afoot to reduce personal tax rates.

The position of farmers is not perfect. Some are doing very well, while others are struggling with difficult world markets. Many would like to see the New Zealand dollar rather weaker than it is at the moment (March 1996). Most farmers, however, are farming successfully without subsidies.

The time it took New Zealand to reach this position was far longer than it should have been, and the time it took to reduce the protection of the manufacturing industries and the work force meant that their level of subsidy was much greater than that received by farmers.

We are fortunate that the National Government, although some of its members were very critical of Federated Farmers' policies while in Opposition, has gone ahead and implemented these policies with great vigour.

The major reformers have received little in the way of thanks for their efforts, however. A number of Labour Ministers lost their seats in the 1990 election. National won the 1993 election only by the skin of its teeth, and several of its MPs have since left the party.

A referendum held in conjunction with the 1993 election resulted in the electorate voting to change the way New Zealand elects its government from 'first past the post' to mixed member proportional representation. It seems that a majority voted for this change because they had lost confidence in politicians.

I believe that National's relatively poor showing in the 1993 election is largely attributable to its failure to deliver on the unrealistic superannuation promise, rather than to its running of the economy. Ruth Richardson was a casualty and did not survive as Minister of Finance.

Hindsight is a wonderful thing. There is no doubt in my mind that we would have been better off if protection had been taken from manufacturers and unions at the same time as subsidies were taken from farmers. Failure to do so has meant that there has had to be much more emphasis on monetary policy and high interest rates to control inflation than there should have been.

On the other hand, it would probably have been politically impossible to reduce the state sector, cut government spending, free up the labour market and reduce manufacturing protection, had farming not been restructured first. Probably only a Labour Government could have dismantled farming support, and only a National Government could have attacked social spending and freed up the labour market.

Given another chance, I am sure that the politicians concerned would handle things differently. This applies particularly to the inflation and deficit blowout which took place during the first period of reform (between l984 and 1987). This made things much more difficult (for farmers especially) than they should have been. As it is, there has been some fallout from both major political parties.

New Zealanders can all be thankful, however, that the reforms have been carried out. We are in much better shape as a country than we would have been if nothing had been done. Doing nothing was, of course, never an option - we were on the point of bankruptcy and had to haul ourselves out somehow.

HOW THE FARMING SECTOR COPED WITH CHANGE

There are just over 80,000 rural land holdings in New Zealand. This number has remained constant since 1984, despite all the changes. There have been both amalgamations and subdivisions. Approximately half of these holdings are commercial farms. The rest are 'lifestyle blocks' or small hobby farms, with occupiers making most of their income from other sources.

As reported in chapter 7, the then Prime Minister David Lange had predicted that up to 8,000 farmers could lose their farms - that is, about 20 percent of commercial farmers in New Zealand would go out of business. This would have been quite unacceptable to the farming community, but in the event it did not happen. Some farmers did leave the land. No-one knows the exact number, but a reasonable estimate seems to be about one-tenth of that predicted by David Lange.

There are many reasons why the prediction was wrong, one being that farmers are resourceful people who can cope in very difficult times. Their initial reaction was to cut spending to the bone. This was partly in response to a Federated Farmers 'shut up your chequebook' campaign, which was instituted in 1984-5 for two purposes. The first was to keep farmers from going further into debt in a time of reduced incomes. The second was that Federated Farmers considered that the flow-on effect of such a campaign would impact on the rest of the community, slowing down the economy and encouraging the government to enact other much-needed reforms.

The campaign was roundly criticised by those in the farm service sector, as their incomes were drastically reduced. Some claimed that the campaign was totally irresponsible. I do not believe this. If farmers had continued to spend at the same rate as in previous years, many of them would soon have been insolvent.

Farmers were also encouraged to make use of supplier competition, which became increasingly intense as the service sector fought to maintain its share of farmers' essential spending. This action certainly reduced prices.

Fertiliser is a large expenditure item on most farms, and its application fell heavily, as Figure 1 shows. Chemical sales also fell.

Livestock numbers were reduced. Sheep numbers in particular fell by 20 million over the reform years (see Figure 4). This was a logical progression in that much of the artificially inspired increased production came from this sector. Much of the expensive to maintain hill country land also ran sheep. Some of the farms had been overstocked anyway, and reducing numbers made sense as farmers could then improve the performance of the remaining animals, which gave a short-term boost to cash flows.

A number of farm workers lost their jobs because farmers could no longer afford to pay them. They were either not replaced at all, or replaced by unpaid workers from the farming family. Many farming women increased their farm work loads at this time.

Farming families also sought off-farm work to supplement the farm income. It was often the women who used their teaching, nursing and other skills to get work locally, but sometimes the men got contracting work for other farmers or professional work in the cities while their wives took over the farm management. People who were able to diversify into off-farm investment did so, though tragically some actually borrowed to do this and lost heavily later.

Interest rates soared soon after the reform process began and life became more and more difficult for many farmers, no matter how hard they and their families worked. Those who had borrowed heavily for development and those who had purchased farms with little equity under the policies of the previous government were hit hardest. Some farmers in the latter category had been encouraged to borrow 90 percent of the ingoing farm price, and many found themselves in an impossible financial position. Those who, in 1984-85, had borrowed off-shore in search of lower interest rates and had subsequently been caught by an appreciating New Zealand dollar were in a similar position.

The package that the government introduced through the Rural Bank in 1986 alleviated the crisis. The Rural Bank was the principal provider of mortgage funds to farming, while trading banks and stock firms were the major source of seasonal finance. In this package, concessional interest rates were increased to market rates in a single step, but at the same time the principal owing was reduced, with loan repayments remaining at the same level. While this did not improve the farmers' cash flows, it did improve their equity. This gave seasonal financiers the confidence to continue lending.

Some 5,000 farmers were assisted by this package, but the problem remained for those financed outside the Rural Bank. Several efforts were made to sell up those who were seriously in arrears to financial institutions. These attempted sales were often disrupted by protesting farmers, who encouraged potential buyers not to bid. A great deal of media attention was focused on these sales, and banks soon shied away from the bad publicity.

The impasse was largely resolved at the 1988 Federated Farmers conference, after which the New Zealand Rural Trust was set up under the chairmanship of Peter Elworthy. Its other two members were Andra Neeley, an emerging farming leader, and David Baker, a farm consultant. The chief executive was a well-known troubleshooter, Jock McKenzie.

The trust employed a team of facilitators throughout the country, their role being to bring borrowers and lenders together in small conferences so that settlements could be negotiated. Either borrower or lender could apply to the trust for assistance, and if a solution could not be found by the parties, the conference chairman's recommendation prevailed. After a number of agreements had been reached, financiers and clients were often able to negotiate without calling in the trust.

The general pattern was that financiers agreed to charge interest rates the farm could afford, provided that the farmer was efficient and the venture would be profitable under normal circumstances. In the relatively few cases where the farm could not be made profitable, ways were found to resettle the farming families elsewhere, without bankruptcy. Most were able to take some money with them.

The concessions made at this time by financial institutions may appear generous, but they were also good business practice. A lot of forced farm sales would have collapsed land prices, with both farmers and bankers becoming big losers. By assisting most farmers through to better times and, with some sensitivity, helping others to leave the industry, the financial sector averted a crisis. Banks actually lost far more money in other sectors of the community than they did in farming.

Two agricultural sectors were affected much more seriously by the changes than others. The first was sheep farming, which had been the major recipient of government support. The second was kiwifruit growing, which had been a glamour industry.

The kiwifruit industry had made spectacular growth on the back of high product prices, government export incentives, and tax write-offs. The price of land suitable for growing kiwfruit went through the roof in the 1970s as people rushed to take part in this bonanza. The land price then fell as the market plummeted with increased production, and the incentives and tax write-offs were removed in the reform process following the 1984 Budget.

The kiwifruit industry is still undergoing change as growers adjust to the fact that their product is now a commodity rather than a luxury item. Most growers favour single-desk selling, and the fruit is now sold through the Kiwifruit Marketing Board. While many efficiencies have been made, prices to farmers continue to disappoint because of the intense competition from other countries.

The plight of sheep farmers was of great concern to the Federated Farmers' Meat and Wool Council, especially after the lamb prices collapsed in 1987. After consultation with the Meat and Wool Boards' Economic Service, it was agreed that the price for an average prime lamb needed to be $30 if the industry was to be sustained. The price was then between $16 and $19. The 'Thirty Dollar Lamb' campaign was born. We believed that improvements could be made to growing, transporting, processing and marketing lamb - and improvement in every sector was necessary if the $30 goal was to be reached.

By this time, ownership of almost all sheepmeat processing and marketing companies was in the hands of farmers. A meeting was called between Federated Farmers, the Meat Producers Board and representatives of the farmer-owned meat companies to discuss the project. Some meat company people considered that the $30 target was unrealistic, but there was general agreement that everyone should work together to get prices up.

Savings were finally made in all the areas targeted. Farmers produced better and heavier lambs. It costs as much to process a small lamb as it does a big one, and a steady increase in average lamb weights has improved efficiency. Better organisation reduced the cost of transport from farm to processor. In the plants themselves large gains have been made through rationalisation, introduction of new technology and reduction in labour costs. Competition in world shipping has been used to maximum advantage. A Meat Industry Planning Council, with representatives from meat companies and the Meat Board, has worked to improve market performance.

The meat industry carried a relatively high debt loading, and the high interest rates prevailing at the time affected prices paid to farmers both directly and indirectly. High interest rates were targeted in a joint campaign by Federated Farmers, manufacturers, Chambers of Commerce and consumer advocates. This group attacked the concentration on interest rates as the major way of controlling inflation, and asked the government to take the other necessary measures.

Government was not seen as the only culprit. Bank margins were also very high, and banks appeared to be using existing clients to pay for losses incurred in the stock market collapse. Loyalty to one bank, particularly from farmers, was traditional in New Zealand. Many people stood in awe of their bank managers, and banks had created the impression that it was a great privilege to borrow money.

Our campaign set out to change all that. Federated Farmers made a video entitled 'How to Deal with your Bank Manager', which was distributed throughout the country. It was our view that if you were creditworthy and had prepared your case well, the bank should welcome your business; and that if one bank did not offer you a reasonable deal, you should definitely try an alternative source of finance.

This campaign received strong public support, and it made national television several times. It is hard to measure its impact but very strong competition has developed among financial institutions for farm business. Packages are now much more innovative, and costs of packages offered are published and compared in the media.

The outcome of the 'Thirty Dollar Lamb' campaign has been that from 1989 to 1995 the average lamb price has been almost $33, peaking at $40.16 in 1993. When the price fell away in 1995, further action was taken to rationalise meat processing (see Chapter 20).

Farmers have often been exhorted to diversify. Some of the new ventures have been successful, while others have disappeared very quickly. During the period of change, diversification was again suggested as a solution to farmers' problems. People were encouraged to get into goats, deer, llamas, ostriches, alpacas, emus, a variety of crops, farm tourism, and even into off-farm investment.

The results have been varied. There is an old saying that those who can afford to diversify, don't; and those who can't, do! There is some truth in this. Farmers who were taken in by the superlatives put about by some people in the goat industry, for example, and who borrowed to buy overpriced stock, suffered very badly. On the other hand, most of the people who stuck with their existing businesses and ran them well came through the change in fairly good shape.

Overall, farmers who diversified cautiously into well-run areas such as deer farming, or who converted a portion of their unused assets into ventures such as farm-stay tourism have done well. Many of those who made speculative investments on and off the farm are among the few hundred who did not survive the changes.

New Zealand farmers have come through the transition from a subsidised industry to the least-subsidised farming industry in the world with very little fallout. The main reasons are:

- Many farmers had very little debt and did not have to cope with high interest rates.
- Many of those who were substantially indebted had their debt restructured. Many also worked off-farm to bring in money to reduce debt and make them less vulnerable.
- Management of most farm businesses was of a high standard.
- Farmers worked collectively through their organisations to bring about the changes required for long-term viability.
- A huge improvement was made in the efficiency of the whole farm service sector.

Many of those in the farm service sector had to make greater changes than farmers themselves did. Previously farming was grossly over-serviced, but the reforms have created a much leaner and more effective service sector. Most of the subsidies to farmers had actually ended up with the service industries. When that money was no longer paid, service industries had to adapt or go out of business.

In retrospect, farmers probably cut back rather too much in vital areas such as fertiliser. This certainly had the short-term effect of bringing income and expenditure into line, but in the medium term it reduced the productive capacity of the land. Those who carried on applying fertiliser, even if they had to borrow to do so, now appear to be in a stronger position than those who did not. At the time, however, farmers often felt that they had no choice.

THE FIGHT FOR FAIR TRADE

No amount of reform can help an exporting country unless it has access to markets. The New Zealand reforms were under way but a bigger battle had just begun, with agriculture included in the Uruguay Round of the General Agreement on Tariffs and Trade (GATT) for the first time. The Cairns Group was set up to fight that battle.

'Who is this Cairns Group, and what has it to do with the GATT negotiations? This is a debate between Europe and the United States, and it has nothing to do with the rest of you.' Said to me by Jan Hinnekens, a European farm leader in 1986, this was an early indication of attitudes in Europe toward the Uruguay Round.

The answer to his question was that there are as many people (and many more farmers) in the Cairns group of countries as there are in Europe and the United States combined. We are a group of countries for whom agricultural exports are vital, and the GATT negotiations had everything to do with us.

In New Zealand's case, if we did not have access to markets we did not have an industry. Being the most efficient farmers in the world would be of no use as long as protectionism and subsidisation in the wealthy industrialised countries continued to grow. As John Kneebone, a former president of Federated Farmers, once put it, 'We would soon become just a quaint little tourist destination.'

The negotiation of any meaningful rules for agriculture had always been too difficult in previous GATT rounds. This is the reason that the Cairns Group (named for the Australian city of its first meeting) was formed. Its members are Canada, Brazil, Colombia, Argentina, Uruguay,

Chile, Fiji, New Zealand, Australia, the Philippines, Malaysia, Thailand, Indonesia and Hungary.

Liberalisation of trade in sectors other than agriculture in earlier GATT rounds had brought great benefits to the industrial world, and Cairns Group members were determined to achieve similar results for agriculture. Paradoxically, the countries that had gained so much from trade liberalisation in other sectors were the strongest opponents of trade liberalisation for agriculture.

New Zealand Federated Farmers was involved with the Cairns Group from the beginning, with Peter Elworthy and Ness Wright at the first meeting. Our major task was to use our contacts in farming and commercial organisations worldwide to gain support for fair trade in agriculture. Federated Farmers was a member of the International Federation of Agricultural Producers (IFAP), the Pacific Basin Economic Council (PBEC), the Japan/New Zealand and Australia/New Zealand Business Councils and the British/New Zealand Trade Council. This was where we began our work.

IFAP is the nearest thing there is to a world farmers' organisation. It has member organisations from most developed countries and a number of emerging nations, but still falls somewhat short of truly representing all farmers. It is, however, recognised by OECD, GATT and the FAO as the voice of the farmers.

Peter Elworthy and I, as President and Senior Vice-President of Federated Farmers, agreed on a division of responsibility which saw his concentration on the home front and mine on the GATT negotiations. So began a campaign which was to occupy much of the next seven years of my life, through my presidency of the Federation and later as a diplomat.

I attended my first IFAP meeting in May 1986, in Bonn. This meeting was a real eye-opener for me. The Australian, New Zealand, Argentinian, South African and Zimbabwean representatives seemed to be the only ones who favoured reform of any magnitude.

The most active Canadian organisations in the IFAP favoured protection. Despite the fact that the United States administration was advocating the elimination of all trade-distorting subsidies, the Farmers Union (the only US organisation active in IFAP at the time of this meeting) took a completely different approach.

The Nordic countries were very keen IFAP members, also favouring protection. Most European farm organisations considered their European organisation, COPA, to be more important to them than IFAP, but they were not backward in using IFAP as a forum in which to argue against fairer trade. Members from developing countries were somewhat constrained in that they often depended on European organisations to pay their way in IFAP.

Despite the odds stacked against us, I think that Ian MacLachlan (Australia), Eduardo Zavaglia (Argentina), David Hasluck (Zimbabwe) and I gave a fair account of ourselves, although we got very frustrated

The author with Belgium's Minister of Agriculture, De Keersmaeker, at Federated Farmers' offices in 1990.

with the chairmanship of IFAP President, Baron von Herriman, a German member of parliament as well as president of the German farming organisation. It seemed to us that von Herriman was using his position as chairman to support those with his own views and to oppose the rest. I challenged him on this during a break in proceedings. His response was that I needed to understand that there was an election coming up in Germany, and he had to be re-elected.

After all of this I was very surprised to find at the end of the meeting that I was to be invited to become a vice-president of IFAP, and that it had been agreed that the agricultural question did need to be resolved in the GATT round. Even at this early stage we had made some progress in convincing the members that export subsidies were totally unfair. As Ian MacLachlan so clearly put it, 'It is impossible for countries with small economies to compete against the treasuries of the United States and Europe.'

IFAP executive members have regular meetings with FAO, OECD and GATT administrators. Those who favoured fairer trade agreed that there would always be at least one of us at any such meeting. As we all came from the southern hemisphere, this was no easy task.

Few northern hemisphere delegates appreciate how difficult and expensive it was for those outside Europe and North America to participate in the Uruguay Round. For most Europeans, visiting the EEC Commission or OECD is like flying from Auckland to Christchurch. North America is about eight hours away. However, for southern hemisphere delegates it can mean 30 hours' travelling each way.

While the original American GATT offer of phasing out all trade-distorting subsidies would have been wonderful for New Zealand, it was never really a runner because of strong opposition from Europe, Japan, and even within the United States itself. The negotiations gradually developed into agreement that there were three areas of trade distortion - border protection, export subsidies, and internal support. Discussion eventually centred on achieving a 30 percent reduction in each area.

There were furious debates about which actions would do most good to agricultural exporters and least harm to recipients of support, what system should be used to measure support and on what date measurements and reductions should start. The EEC, for example, claimed that it had already reduced support before negotiations started, and this reduction should be taken into account retrospectively.

Increased market access was the most feared change, as people believed that their internal regimes would not survive the competition from imports. Many of the IFAP countries protected their local markets in the name of food security. Prices paid to achieve this led to over-production and a need to dispose of surpluses. To stop over-production, supply management systems with controls such as quotas and set-aside were instituted.

These systems can be a lucrative proposition for those lucky enough to be involved in them, but they make entry into the industry very difficult. In the UK, for example, anyone entering dairy farming has to buy a quota, which can cost hundreds of thousands of pounds, on top of buying the land and the cows. There are also problems with high food prices and less consumer choice.

Despite the problems with supply management, many IFAP members advanced it as an alternative to more liberal trade. The so-called advantage to those of us who depended on exports was that we would then be able to export to developing world markets without 'undue' competition from those practising supply management. The Nordic countries, Japan, many European farmer groups and some US and Canadian organisations were strong supporters of this idea.

These Canadian organisations were a great embarrassment to the Cairns Group. Canada had been accepted into the Cairns Group because Canadian grain exporters and beef farmers were oriented toward fairer trade, and it was felt that this attitude would ultimately prevail over the supply management lobby in the dairy and chicken industries. In the end it did not, and the supply management people had the resources to fight change in forums all over the world.

I well remember one meeting with GATT in Geneva where most countries were represented by one or two people. Canada had about 40 people in attendance, most opposed to reform. I also remember New Zealand's chief negotiator, Tim Groser, at the abortive 'final' GATT meeting in Brussels in 1990, wearing a 'Support Fairer Trade' button on one lapel and a 'Supply Management' button on the other. When asked if he was schizophrenic, he joked, 'No, I'm Canadian!'

During the GATT negotiations it was common for New Zealanders to be asked to speak at conferences all over the world. We were the only nation to have dismantled a subsidy system, and we had a credibility in the debate that others could not match.

Much of our work was reassurance. There was a perception that New Zealand was a farming paradise in the South Pacific, waiting to flood the world with cheap food. This is of course not possible. We are just a drop in the bucket in world production terms (eg, less than 2 percent of world dairy production) and collapsing world prices would certainly not be in our interest.

On the other hand, we had to point out that continuation of subsidies by countries that were real forces in world production terms would drive countries like ours to the wall. Our situation was bad enough, but the plight of countries that had been exporters of cane sugar before the subsidisation and protection of sugar beet and corn syrup in Europe and North America was simply tragic.

Two speaking highlights for me were the Oxford Farming Conference in the UK in January 1988, and the World Food Conference in Brussels in April of the same year. Oxford was a highlight because a large audience in a country with high subsidy levels gave me a very good hearing. I made friends at this conference who have travelled the fair trade road with me ever since.

The World Food Conference was chaired by Lord Plumb (formerly Sir Henry), then President of the European Parliament, and former President of the National Farmers Union of England and Wales. The conference was called to try to resolve the dual problems of food surpluses in the Western world and Third World hunger. It had been thought that the first could solve the second, but as both continued to increase, this was obviously untrue.

Kenneth Kaunda, President of Zambia, captured the conference early on the first day with the simple statement that the Third World needed food trade, not food aid. I was able to build on this, and to show that it was possible to farm without subsidies in a developed country. In his summing up, Lord Plumb said that 'the subsidy madness' had to stop. I like to think that my presentation of the New Zealand experience had some influence on his conclusions.

The support for change gradually gained momentum. The return to IFAP of the American Farm Bureau, the largest farming organisation in the world, brought its President, Dean Kleckner, a most articulate supporter of trade liberalisation, into the arena. This helped to bring balance back into the IFAP debates.

The 1988 IFAP Congress in Adelaide produced a surprisingly even-handed communique on GATT. Rob McLagan, chief executive of Federated Farmers of New Zealand, was on the drafting committee, and I have no doubt that he used all his skills to help achieve this result. It was ageed, however, that IFAP should call a conference on supply management so it could be debated as an alternative to what was being promoted in GATT circles.

This conference was held in Spain in 1989. I was asked to present a paper on 'The Market as an Alternative to Supply Management'. My comments upset many Canadians and others, but I was heartened and a little surprised to receive strong support from some of the Spanish growers we met.

My colleagues and I decided that I should run for the IFAP presidency in 1990. We believed that it was time for a president who represented a wider group than the wealthy northern hemisphere countries, and who believed in fair trade. The majority of Europeans had decided months before that Hans Kjeldsen of Denmark would be their choice, and they worked very hard to keep me out. The final GATT meeting was set for December 1990, and they saw it as crucial that a European led the world farmers into that meeting. Anything else would send what they believed was the wrong signal to the GATT meetings.

There were minor hitches to their plans when other candidates popped up, and when the African countries began supporting me strongly. One African delegate told me that God had chosen me to lead the farmers. The Swedes tried to defuse the situation by suggesting that the current president, Glen Flaten of Canada, stay on an extra year.

I knew that I had lost the election the night before the vote, when I was summoned to a meeting of African leaders and asked what I could offer them as president. I spoke about strong leadership, especially in trade issues. I was then told that this was not enough. I would have to match aid offers that had been promised to them in return for their support. Afterwards, the standing joke around New Zealand farming circles was that by losing the election, I had got more aid into Africa than I could have done in ten years by winning it!

This election was also a good example of how Britain now lines up with Europe ahead of the Commonwealth, most of which was on the fair trade side of the argument. When he wished me luck for the election, British High Commissioner in New Zealand, Robin Byatt, was very surprised when I told him that the National Farmers Union of England and Wales would not be supporting me. I do not hold this against my British friends - Europe is simply much more important to them than the Commonwealth is.

Hans Kjeldsen won the election on the first ballot. Some of my African friends were really upset on my behalf, especially the one who felt that he knew God's plans for me. I told him that God must have something else in mind for me, and he cheered up considerably.

Amazingly, religion played quite a part in this election. A European farming colleague, deeply embarrassed by his role in this, confessed later that as a Catholic he had been deputed to win the African Catholic vote, with much being made of the fact that I was a Protestant. That Kjeldsen was also a Protestant was to be kept a secret.

In hindsight, all the efforts that went into making sure I did not win backfired. Within IFAP, my hands would have been tied. Outside its confines, I could campaign wholeheartedly for fairer trade and more particularly for New Zealand's case.

One door had closed, but a much more rewarding one had opened.

MISSION ALMOST ACCOMPLISHED

Agricultural protectionism and subsidisation had become so entrenched in a number of countries that those receiving the benefits considered them to be their right. I am also sure that most of the farmers getting the subsidies had no idea of the damage they were doing to the livelihood of fellow farmers in other countries.

The farmers did not seem to know that their countries had become wealthy because previous GATT rounds had freed up trade in the industrial sector, and this wealth was being used to protect them against competition in agriculture. As long as agricultural trade did not come under the same rules as industrial trade, they could keep their privileged position and poor farming nations would stay poor.

Most of the farming families in these wealthy countries are not particularly wealthy themselves. On average they are hard-working, solid citizens. The big landowners, however, have become very affluent and hold the positions of privilege by world standards.

In most instances, and particularly in the European Union, the farming leadership has come from the privileged end of the farming spectrum. As these people have been anxious to hold on to their advantage, they have gone about convincing all farmers that their survival depends on the continuation of the system they now have. It could be said that instilling a fear of change into the minds of their members has been their greatest achievement.

All sorts of dire predictions have been made and believed, and farming leaders have linked up with representatives of other protected industries, such as the United States textile unions, to oppose any change.

Probably their greatest public relations success was the very vigorous campaign to convince the numerous aid and environmental organisations of the world that a GATT settlement including agriculture would destroy the environment and make the rich richer and the poor poorer. Dealing with the misinformation and just plain ignorance was to be a huge task for those of us favouring reform.

The fairer trade side of the battle had much less in the way of resources and personnel. Despite the population of the Cairns Group countries and the number of farmers in this population, most of these people were very poor by comparison with the farmers of Europe, Japan and North America. When I mentioned to the Thai Minister of Agriculture that it would have been of great help to have some of his farm leaders supporting the work of the Cairns Group in other countries, he said simply that they were too poor to be able to travel. As the Canadians could not agree on a united view, it was largely left to Australia, New Zealand and Argentina to present the Cairns Group case.

I entered the fray again as a diplomat. Mike Moore, New Zealand's Minister of Overseas Trade, asked me to take up an appointment as Special Agricultural Trade Envoy when I completed my term as President of Federated Farmers in 1990. This appointment meant devoting myself full-time to campaigning for a GATT settlement including agriculture. My work was funded jointly by the government and the farming sector (Federated Farmers and the producer boards), and carried ambassadorial status.

The ambassadorial status was an interesting innovation on Mike Moore's part but was, as he put it, 'the status without the salary'. It did serve, however, to get me interviews with high-ranking people. The appointment was exactly what was needed at the time. It was uniquely suited to my areas of expertise and to the Cairns Group cause, and since I was free to travel wherever I was needed, it became a very effective mission.

The New Zealand Ministries of Foreign Affairs and Agriculture supported my work at all levels, and I was able to carry out an extensive campaign in many countries. When the government changed in November 1990, Philip Burdon was appointed Minister for Trade Negotiations. In 1991 he invited me to continue the work.

New Zealand had a special role to play because it had dismantled its subsidy system. We were able to say from our own experience that many of the dire predictions about subsidy withdrawal had not happened to us, and were unlikely to happen to others.

People from all walks of life had a genuine interest in what we had to say. Through personal meetings with politicians (I met hundreds of them), business leaders, bureaucrats, and people in consumer and charitable organisations, as well as through writing and distributing articles and fact sheets in many countries and in many languages and also taking part in interviews on radio and television all over the world, I and others working alongside me were able to influence millions of people.

My initial campaign as Special Agicultural Trade Envoy in 1990 was to fire up the supporters - or those we considered should have been supporters - of a settlement in the Uruguay Round involving agriculture. The second push, in 1991, involved stepping into the opposition camps and debating the issues logically and sensibly.

My later work as Counsellor (Agriculture) at the New Zealand High Commission in London, while principally set up to defend New Zealand's access for sheep-meat and butter to the European market, also involved a great deal of campaigning for a GATT settlement as the Uruguay Round ran far over its allotted span.

Throughout the whole period from 1984 to 1993 when a deal was finally struck, the challenge was daunting. Beating back agricultural

On the diplomatic circuit in Wales, 1992, at the farm of Andrew Jones, Welsh president of the National Sheep Association: (l to r) Andrew Jones, Twnog Davies (Regional Socio-Economic Adviser), His Excellency George Gair (NZ High Commissioner to the UK) and Fay Gair, and the author.

protectionism in many countries is something for which few politicians have the courage. In some countries, the farming lobby has a strength far beyond the number of people involved in farming. Japan is a good example. Rural votes are worth much more than urban votes, and can be used to give farmers the balance of power, which has then been used to provide Japanese farmers with such benefits as rice payments that are seven times greater than world prices. In Europe the Common Agricultural Policy is entrenched. Massive demonstrations are often mounted if there is any suggestion of change.

In New Zealand at the time, there was quite a lot of pessimism about the fair trade lobby's likelihood of success in the GATT negotiations. Even Jim Bolger, then Leader of the Opposition and whom I regard as an optimist, told me in July 1990 that I was wasting my time as the Europeans would never change their stance.

However, we did have a few things going for us. Chief of these was probably the calibre of the key players in the GATT negotiations. I remember particularly Arthur Dunkel from Switzerland, Director-General of GATT for most of the process, who despite the protests of Swiss farmers never gave up trying to achieve a settlement, no matter how difficult it became.

There was also Aart de Zeeuw, the likeable Dutchman who was the initial chairman of the Agriculture Negotiating Committee. The final settlement in agriculture was not very different from a paper he produced in 1990.

Others who had a high profile included Irishman Ray MacSharry, who was Agriculture Commissioner for the EEC for much of the time, and who drove the reform of the Common Agriculture Policy. Clayton Yeutter held two key roles. Initially he was US Trade Secretary, and later Secretary of Agriculture. He was followed as Secretary of Trade by Carla Hills. Although they had all moved on by the time the tortuous process was finally completed, in my view these were the key people in shaping the agricultural settlement.

New Zealand had three chief negotiators over the period: Ted Woodfield, Richard Nottage and Tim Groser. These men did great service for their country, and all New Zealanders should be very proud of them and their staff. George Rutherford and his colleagues from the Ministry of Agriculture also did sterling work.

I will never forget my first meeting with Ray MacSharry. He told me that where he came from in Ireland, the average dairy herd was sixteen cows. It had been that way for many years, and if he had anything to do with it, it would stay that way. That is what he had come to Brussels to preserve. His reform of the CAP in 1992, while considered fairly modest in New Zealand, was attacked from all quarters in European farming. The wealthy farmers were strongly opposed, and even those with sixteen-cow herds were unimpressed. Alan Gillis, then President of the Irish Farmers Association, told me that they would 'spill blood' to stop the reform going ahead. To MacSharry's credit, he stuck to his guns, and this helped move the GATT process forward.

As New Zealand's politicians have found out, being a reformer is not always easy nor is it always appreciated by your electorate. Swedish Agriculture Minister Matts Helstrom chaired the Agriculture Negotiating Committee at the 1990 meeting in Brussels. He put up a compromise solution, which proposed that a 30 percent reduction in support be made in each of the trade-distorting sectors. Despite his efforts, the meeting failed, mostly because of the intractability of the Europeans. When I suggested to the leader of the Swedish farm organisation that their Minister had done a great job, he did not share my view.

'Wait till we get him home,' he said. Sure enough, Minister Helstrom's party was defeated at the next election.

British Minister of Agriculture, John Gummer, was one politician I could not warm to. Perhaps it was because he was prone to making sweeping criticisms of New Zealand, especially when we were sharing a platform. Once he said that, unlike the British Government, the New Zealand Government did not care what happened to its farmers. He went on to say that agriculture was more important to Britain than it was to Australia or to Canada. I have no idea on what statistics he was basing this statement, but it is, of course, complete rubbish.

There were many frustrations and disappointments throughout this campaign, but on occasion there were highlights. I received a great boost when the Deputy Prime Minister of Canada told me, on the eve of a meeting we were both to address, that he really liked my speech. As I had not distributed my notes at that stage, I wondered what he was talking about. It turned out that he was referring to a speech I had made to the Press Club in Washington DC a few days earlier, which had been sent on to him by Clayton Yeutter. I found out later that Yeutter had been sending my speeches to a number of people around the world.

When I caught up with him a few weeks later, I mentioned this and thanked him for helping with my campaign. He said that he had done what he had because they were fine pieces of work for the cause. All of which goes to show that a New Zealand voice can have some influence in international circles.

Many people outside officialdom have also campaigned with great persistence for trade reform. I hesitate to mention names, but must make an exception for Anthony Rosen, a British journalist and consultant. Ever since I met him at Oxford in 1988, I have been impressed by the way that Anthony has fearlessly promoted fairer trade, despite the criticism he frequently gets from within his own country.

Lord Plumb has also made a very positive contribution. His chairmanship of the International Policy Council on Agriculture and Trade, formed in 1988, has played an important part in the process.

The final settlement of the Uruguay Round was somewhat of an anticlimax in a negotiation that could have delivered much more. Many New Zealand farmers were disappointed, especially as it will be the year 2000 before the improvements reducing subsidisation and protection are finally in place.

The agreement is, however, a good start, and we are already way ahead of where we were. An increasing number of farming leaders in countries like the UK, Denmark and the Netherlands now see the advantages of more liberal agricultural trade and are starting to campaign for it. We have to remind ourselves of where we would be now if there

The author on the day the GATT deal was signed. Photo courtesy of the New Zealand Herald.

had been no agreement, and protectionism and subsidisation had continued to grow unchecked.

I doubt that we have seen the end of reform. Budget deficits and a new understanding of the benefits of reform appear to be hastening change in the United States. Accommodating the Central European countries in the EU on a budget which will decline in real terms could mean further reform of the CAP.

Key Asian and Pacific nations, including the USA and Japan, have formed the Asian Pacific Economic Council (APEC). Its leaders have agreed to eliminate all trade barriers by the year 2020. We are on the way.

The Uruguay Round has undoubtedly been the catalyst for all of this change, and the Cairns Group played a vital part in making it happen. A settlement on financial services and intellectual property was crucial to the industrialised countries. The Cairns Group was able to say, 'No deal on agriculture - no deal on the rest of the items on the negotiating table.'

A small band of dedicated New Zealanders was at the heart of the Cairns Group campaign, and did more than its share of the huge amount of work involved. Those concerned came from government ministries, Federated Farmers, and the producer boards. They were supported by both Labour and National politicians, and led enthusiastically by Mike Moore and later Philip Burdon, as trade ministers.

The final settlement of the Uruguay Round came with little fanfare on 15 December 1993 - three years after its due date. I was not at that time doing any official work on trade reform and the day, while highly significant, seemed somewhat unreal. My wife Jan, who had shared all the highs and lows of the GATT campaign with me, marked the occasion by designing and embroidering a sampler for me, choosing for the centre a quotation from one-time US President Calvin Coolidge: 'Persistence and determination alone are omnipotent.' This sums up the efforts made in the 'David and Goliath' fight for fairer trade. I am proud to have been part of it.

THE MYTHS OF DESERTIFICATION AND POLLUTION

In 1992, the Duke of Edinburgh told me that Europe would become a desert if it practised New Zealand's policies on unsubsidised farming. He seemed to be serious at the time.

I should not have been so surprised. Prince Charles has been known for his pro-subsidy beliefs for many years, on the grounds that removal of subsidies would ruin both the country way of life and the environment, and I suppose it is a case of 'like father, like son'.

An interesting footnote to this is that the Duke of Edinburgh, on his 1995 visit to New Zealand, asked the Ministry for the Environment about the effect on the environment of the removal of farm subsidies. Perhaps he is reconsidering his opinions on the subject.

The Duke was, in 1992, expressing a view widely held in Europe. The French, in particular, have made great use of the jargon word 'desertification' in their opposition to any alteration to the Common Agricultural Policy, especially through the medium of the General Agreement on Tariffs and Trade.

Public relations campaigns run by farm organisations and farm servicing industries that benefit from the subsidies seem to have convinced many Europeans that desertification is a real possibility - when it is just a myth. It is completely absurd to suggest that some of the best farming land in the world, centrally situated in one of the most affluent markets of all, would not be used for food production if subsidies were not paid. Even if the land were allowed to return to its natural state, it would not become a desert.

Although few people try to argue that the best land would not be used if subsidies were removed, there are plenty who claim that areas of poorer soil, harsh climate or steep terrain would turn to desert. They say that rock walls, grassy hillsides, neat hedgerows, flower gardens and tree plantings are all there because of subsidisation of farm produce, and imply that they would disappear as soon as the money stopped coming in.

In an effort to keep the poorer land in Europe in farming, two special land categories have been created: less favoured and severely disadvantaged. These areas receive extra subsidy support on top of the basic production subsidies, border protection and export subsidies that are the cornerstones of the Common Agricultural Policy. It is claimed without this support much of the land would be abandoned, becoming the 'desert' mentioned before.

It is even claimed that there would be no grouse on the moors. It was explained to me while I was visiting a farm in Scotland that if sheep farming was not subsidised, there would be no sheep on the moorlands to attract ticks from the grouse, and the grouse would die of tick infestation!

The New Zealand view of what constitutes a good environment is fairly simple - clean water, clean air and no erosion. On the basis of these criteria, the removal of subsidies in New Zealand has improved the environment.

For many years we had subsidies on the price of fertilisers and agricultural chemicals, and farmers tended to use what was subsidised, regardless of need. Fertiliser was subsidised but fencing materials were not and farmers, being the perverse creatures that they are, would use the fertiliser, even if the money would have been better spent on fencing to make more efficient use of the grass already grown.

Removing subsidies at first reduced the use of chemicals and artificial fertilisers. As farming has recovered, the use of fertilisers has increased somewhat, but the use of both fertilisers and chemicals now has to be justified, and is definitely more specific than it once was. By and large we no longer farm the erosion-prone, fragile land that was developed with encouragement from subsidies. With hindsight, we can see that such land should never have been developed, and it is now being used for commercial forestry or the regeneration of native forest.

I have no doubt that some of the more extreme supporters of the Common Agricultural Policy would dispute my use of the term 'regenerate', and insist that our land is actually becoming desertified. Yet I cannot imagine anything less like a desert than land covered with native trees and ferns. Native vegetation grows really quickly in New Zealand so regeneration does not take long. I have viewed similar regrowth in the United States where such trees are improving the environment in several ways.

The question of pollution and the destruction of the physical environment as a consequence of the removal of subsidies was a hot topic at the time of the Uruguay Round of the GATT. There was a division of opinion, even among Europeans. The Bruntland report, prepared for the United Nations in 1990 by a committee chaired by the prominent Norwegian politician, Gro Harlem Bruntland, made it abundantly clear that over subsidisation and over-intensified production in many countries had had a very damaging effect on the environment. It had caused pollution of land and waterways, and was creating huge problems for those who had to clean up the mess and plan for adequate disposal of surplus effluent in the future.

Prince Claus of the Netherlands, in his opening speech to the Sixth Congress of the European Association of Agricultural Economists in The Hague, 3 September 1990, said:

"We now know that agricultural activities can harm the environment. In the industrialised world these problems are mainly the result of intensive farming methods, which involve the use of large amounts of pesticides and fertilisers. In the Netherlands, for instance, intensive livestock farming has created the problem of how to dispose of vast quantities of surplus manure. Inadequate disposal is causing enormous damage to vital agricultural resources such as soil, water and air."

On the other hand, Jonathon Porritt, former director of 'Friends of the Earth' and a prominent and typical European 'green', told the 1990 conference of RURAL (Responsible Use of Resources in Agriculture and on the Land), that a successful GATT round would devastate small farms around the world. This would lead to bigger and bigger farms, on which there would be little or no interest in the environment or in wildlife.

He was in favour, he said, of integrated agriculture which would produce the kind of food and the kind of countryside which he and many others wanted. He claimed that there were millions of people who, like himself, were prepared to pay a significant sum, either through the price of food or through taxes, to finance such a system.

He emphasised that it was essential that farmers should be able to continue producing food, even if they did so less intensively, and it was no less important that rural communities should remain vital and alive.

Jonathon Porritt should perhaps have thought through these arguments more carefully before taking sides in the GATT/fairer trade battle. The market forces that he portrays as on the side of evil in this debate would be just the allies he needed in producing the food he and 'millions like him' want at a higher price - if indeed there are millions wanting it.

The trouble is that he appears to be so determined that market forces are evil that he advocates others paying for his preferences through higher prices for all food, taxes, or through continuation of subsidies.

The worst part of this argument, and to give him the benefit of the doubt, I will assume that he is ignorant of the facts, is that for subsidies to work, they have to be combined with border protection, to keep out produce from other countries. He is therefore advocating denying access to other countries, including those from the Third World which desperately need this access, in order to get what he wants. He is asking the rest of the world's farmers, almost all poorer than those of his own country, to pay for the upkeep of his ideal farming and rural community.

He then says that farming less intensively would be acceptable. This is where he finds himself actually on the other side of the fence. The result of removing subsidies is often cell intensive farming, especially of fragile land, which is beneficial to the environment.

My proof does not lie in theory, but in practice. New Zealand, the only country in the world in recent years to have built up a system of high subsidies and then to have removed them, has found that to be the case.

Since the removal of subsidies we have seen a much better balance between supply and demand, less farming of marginal and fragile land, a reduction in the overall use of chemicals and fertiliser, the return of much marginal land to forest, natural or commercial, and much more emphasis on cost control and quality rather than quantity in our produce. All of this would probably be exactly what Jonathon Porritt and his friends would want.

The largest food processing company in New Zealand, Heinz Watties, announced in 1995 that it was increasing the number of farmers from whom it is buying organically grown produce. This is just one of a number of examples where people in New Zealand are producing for a particular market, without subsidies, in response to demand. If there is a demand for organically produced food, or indeed any other type of food, market forces will certainly meet it.

When we lived in London in 1992, I was shown a questionnaire that had been used in parts of England to determine people's preferences as to how the countryside should look. Participants were asked to study a series of six or eight artist's impressions of the same country scene, with changes from picture to picture. The first was of a completely manicured landscape with close-cropped grass, animals grazing, tidy stone walls and neat hedges. The last was of a wild scene with no animals, unkempt grass, untrimmed hedges and so on, while the pictures in between illustrated the progression from the first scene to the last.

The participants were asked to say which type of countryside they would prefer to see in England. I need hardly say that most people voted for either the first or the second picture, and the instigators of the questionnaire were using this as evidence that the British people wanted to keep the rural environment in this state, and would pay for it.

On investigating further, I found that the respondents to the questionnaire had not been told how much it would cost them to maintain each rural scene or, more practically, what they would have to forego in terms of government expenditure on other items like health or education in order to fund their environmental dreams - so it was hardly a valid conclusion.

I also realised that Europeans generally had a totally different concept of the word 'environment' from mine. Their interpretation went beyond the physical things that I had been thinking about, such as water and air, and into the social, the cultural and the aesthetic aspects. From then on, I had to think in these terms whenever I heard the environmental argument.

THE MYTH ABOUT DESTROYING THE SOCIAL ENVIRONMENT

While visiting New Zealand in 1995, the Irish Minister of Agriculture, Ivan Yates, argued that New Zealand did not really want an unsubsidised farming industry in Europe.

He painted a picture of economic pasture-based dairy industries in the UK and Ireland, and highly productive grain farming in continental Europe, with larger and more efficient farms. He believed that these farmers would be extremely competitive internationally, to the detriment of New Zealand. However, this situation would not, he said, be allowed to develop, as it would mean that six million farmers would have to leave the land, and the social environment in rural Europe would be left in chaos.

It is this social environment that the Common Agriculture Policy and other regimes like it around the world are meant to protect. Such policies have created huge distortions, both locally and internationally, as efforts have been made to curb production on good land and boost it on the more difficult country. The absurdity of it all was brought home to me when I visited a farm in Scotland with a group of agricultural diplomats in 1992.

This farm had originally been a beautifully balanced livestock unit, with easy rolling land in the front and higher moorland at the back of the property. Grain subsidies came along and the better land was turned over to grain farming, while the livestock was consigned to the poorer land at the back. Not all the better land was particularly suited to grain growing, however, and when set-aside arrived, with payments of £70 per acre for doing absolutely nothing with the land, this area was taken out of production. It then grew beautiful grass and clover that could not be used under the set-aside rules.

At the same time, hay and grain had to be bought in to feed the hungry (but heavily subsidised) stock on the hill section of the farm. The subsidies for the sheep and cattle on the hard hill country were being paid because, and I quote, 'You can't do anything else with the land.'

These policies are meant to keep people in rural areas, keep the fields green, the hedges trimmed and the buildings nicely maintained and cleaned so the countryside will be attractive for tourists and city-dwellers on weekend outings. The philosophy is that if you subsidise the farmers, you will retain a vibrant rural economy.

John Cameron, former president of the National Farmers Union of Scotland, has been one of the strongest advocates of the subsidy system. He defends subsidies in forums all over the world, claiming that they keep people living in the hills and glens of Scotland.

He uses the same arguments as those put forward by Robert Forster (see chapter 5), claiming that sheepkeeping keeps civilisation going in that part of the world: if the sheep were not there, you would see an exodus to the cities, where unemployment would increase and more social services would have to be provided at considerable cost to the government. He argues that subsidising sheep farming actually provides value for money for British taxpayers.

I have seen many examples of this idea of feeding money into farming 'to benefit the rest of the economy'. Norway, for example, probably has the highest ratio of combine harvesters to acres of grain planted of any country in the world. It is not uncommon to see a combine worth more than $NZ200,000 standing in the barn of a 30-acre (12-hectare) farm. It is said that as only a few days in the year are suitable for harvesting, all farmers, even those with small areas of grain - need their own combines to ensure that the crop is harvested, irrespective of the cost. Their high subsidy level enables them to do this.

To encourage sheep farming and to keep the rural areas populated in Norway, the government ensures that farmers are paid a price for wool that covers the cost of production. When I visited Norway in 1990, the price paid for wool was six times the world price.

In Switzerland, huge demonstrations were mounted against a GATT settlement, because the Swiss claimed that they were already losing three farmers per day under current policies. They believed that a GATT settlement would hasten this process and completely devastate their rural way of life.

The Finnish people wanted their border with the Soviet Union kept under surveillance. They therefore paid farmers highly subsidised prices to milk cows in the area, even though it was one of the most difficult dairy farming regions in the world.

In Wales, it is said that a market-driven agriculture would destroy the Welsh language because there would be fewer Welsh-speaking sheep farmers.

Sheep farming in Europe was one of the later forms of farming to be brought into the Common Agricultural Policy. The industry was actually going along quite nicely until officials were persuaded that a regime was necessary, especially to assist French sheep farmers in poorer areas.

In the United States, the supporters of subsidies say that they are necessary to save the family farm and to stop a multinational takeover of agriculture.

Some sectors of Canadian agriculture argue for heavy protection so that they can compete with the United States.

In Japan, rice has special cultural significance, and rice farming also conserves vital water supplies in the system, so rice farmers are protected. The list goes on.

It is appropriate that the results of these schemes are examined. Whether they have achieved their goals is highly debatable, and even where there seems to have been some achievement, the cost has always been borne by others. Consumers and taxpayers in the countries where farming subsidies are paid are obviously big losers, but this is only part of the equation. Those who have paid most are the efficient producers in the rest of the world.

What John Cameron fails to take into account in his championship of subsidisation of hill country farmers in Scotland is the effect that these policies have on sheep farmers in other countries. If the market for sheep meat and wool is not expanded, the subsidy-inspired Scottish production will simply displace another country's produce and almost inevitably reduce the price for all other exporters.

Reduced prices often make very little difference to the subsidised farmer, whose income is usually guaranteed anyway, but they do have a devastating effect on farmers who are solely dependent on the market for their incomes.

In the case of sheep meat and wool, it is usually farmers from New Zealand, Australia or even Argentina whose produce is displaced or reduced in value, and rural communities in these countries pay the price. It is not an exaggeration to say that for every rural school kept open by European sheep subsidies in Scotland, Wales or France, one is closed in a rural area somewhere else in the world.

Let us look at some of the other myths, starting with the Norwegian situation. We were told that the government was paying high prices for grain and wool so that farmers could enjoy a standard of living comparable to that of their city counterparts. They were apparently meant to spend the money on their family's health and education. Instead, to the consternation of officials, it was more often spent on expensive and rarely used farm machinery. This of course might have helped the city machinery businesses for a while, but it did nothing for the social environment in the rural areas.

I had imagined that with the price of wool to farmers being six times the world price, the country people would have used this income to set up their own spinning and hand-knitting ventures, as Nordic sweaters are popular with tourists and locals alike. However, giving them such a huge subsidy on the wool price seems to have eliminated any incentive to add value to their product. It was simply dumped on the world market, thus reducing the price to farmers in other parts of the world.

The Swiss believe that they must protect their rural communities from further loss of farm jobs, and do so by border protection to keep out imports. They justify this on rather spurious social grounds, while ignoring the fact that Swiss industrialists are big exporters and expect the rest of the world's markets to be open to them.

I lost any sympathy I may have had for the Swiss farmers when I watched them arriving in Brussels to protest at the possibility of a GATT settlement. They parked their opulent cars, pulled on their designer leather jackets, lit their expensive cigars, and then tried to tell others that their industry would be devastated if farmers from other countries were allowed a modicum of access to their local market.

To be fair to the Finns, they probably didn't realise that, by introducing to the market produce which would not exist if it were not so heavily subsidised, they were really making rest of the world's farmers pay for their local defence. But indeed this was exactly what they were doing.

Some of the strongest supporters of the Welsh language truly believe that their language will die if sheep products are not subsidised. New Zealand lamb going into the UK is seen as the devil incarnate, displacing Welsh lamb from the market.

Each time I visit Wales, the language seems to be making a resurgence, just as the Maori language is doing in New Zealand. Since the two things are happening at the same time under totally different farming regimes, I venture to suggest that the subsidisation of agricultural produce has less to do with the survival of a language than my Welsh friends may think. If people want a language to survive, they will pass it on to their children, whether they are city dwellers or farmers. They will encourage the setting up of schools that teach in the language, and lobby their local and central governments for funds. Farming, and more particularly New Zealand lamb, should not be part of this debate at all.

Japan has protected its farmers by convincing the rest of the population that they must pay high prices for rice. This is because rice grown in their own country is seen as superior to that grown elsewhere, and because farmers growing rice also conserve valuable water supplies. Of course, to make this policy work, the Japanese government has had to keep out imports from other countries, at a cost to farmers around the world. Japanese farmers have been protected, but the cost to the average Japanese citizen is huge.

The Canadians and Americans (or at least some sectors of their farming communities) like to think that subsidisation is keeping the family farm unit intact. Over the period of heavy farm subsidies, Canada has lost more than two-thirds of its dairy farming families and a similar loss has occurred in Denmark, which has been a keen supporter of farm subsidies. Most subsidy money is, of course, captured by the bigger farmers, who eventually take over the smaller ones.

By contrast, New Zealand rural family figures have been remarkably stable over the years. In 1971, the proportion of the New Zealand population designated rural stood at 16 percent, and it has remained basically unchanged since. To take actual numbers, rather than percentages, the number of people living in rural areas has remained relatively constant at half a million since 1921.

Logic would insist that rural population trends in New Zealand will remain fairly stable in the future, while those in the heavily subsidised countries are likely to continue to fall, if each keeps to the path it is now following.

It is a myth to suggest that if subsidies were removed, no-one would live in the countryside. Many people dream of living in the country and do so the moment they can afford to. They are fully prepared to pay for the experience themselves. In some countries these people are seen as rank amateurs, and local governments try to curb their activities, even when they keep the land producing and employ others to help them.

In Wales recently I met a man who was exhibiting sheep at a show. He told me that his move from a northern industrial city in England to a small farm in west Wales was like 'dying and waking up in heaven'. Nobody had paid him a subsidy to move to the country, and he planned to be a countryman for ever.

Every country in the world has probably experienced an urban population drift as a result of better communications and new technology. Today we are experiencing the beginning of a rural drift, as people take advantage of new technology to work from home in beautiful rural surroundings, rather than in city office buildings. It is a great pity that some governments have complicated life for us all by overreacting to the urban drift with heavy-handed applications of subsidies and bureaucracy.

How much simpler and cheaper it would have been if the Finnish Ministry of Defence had used defence funds to pay a few people to watch the Russian border, instead of paying farmers in that region large sums of money to produce their farm goods.

How much neater it would have been if the Japanese farmers had billed the water users directly, and the Welsh language enthusiasts had paid for their schools from educational grants and contributions from interested parents rather than from farm subsidies.

How much more logical it would have been if the European farmers had been paid directly from, say, the ministries of the environment for cutting the hedges, mending the stone walls and keeping the countryside attractive. Such a system would distort markets much less than paying subsidies on farm produce does.

Hindsight is a wonderful thing, but the fact remains that the New Zealand experience shows that subsidies do little to promote the rural society, and the spin-offs from them do great damage to town/country relationships, to rural societies in other countries, and particularly to the Third World.

THE MYTH OF LOWER PRICES FOR LAND AND STOCK

Two major myths surround land prices and the removal of farm subsidies. The first is that if subsidies are removed, land prices will collapse. The second is that if land prices fall, it will be disastrous for farming.

There seems to be little doubt that subsidies are factored into the price of agricultural land as they are increased over a period. The fear of a collapse in land (and to a lesser extent, livestock) prices is one of the reasons why people resist the withdrawal of support. It is not only farmers who fear this. Financiers with large rural investment are probably even more concerned.

New Zealand experienced this land price rise in the era of subsidisation. In the ten-year period 1972-82, agricultural land prices in New Zealand rose by 574 percent, an inordinately high increase even considering the high inflation rate that prevailed at the time.

Land prices did fall for a time after subsidies began to be removed. However, they did not plummet, as people often predict. In the six years 1982-88, prices fell by 22 percent only. The more interesting progression in prices has taken place in the six years following 1988, when farm subsidy removal was followed by other necessary economic reforms. In this period (1988-94) farm prices rose by 100 percent, so that in 1995 the price of most categories of farm land was higher than it ever had been.

The rate of inflation has certainly affected the rise or fall of land prices in real terms. It can be argued that the 22 percent fall in the six years from 1982 was greater in real terms because it coincided with a period of high inflation. On the other hand, the period since 1988 has seen a large increase in land prices at a time when inflation has steadily fallen, stabilising at a very low level.

Many issues affect agricultural land prices. Some of them are directly related to farming - others are not. For instance, approximately half the rural land holdings in New Zealand are not commercial farms. Most of these are hobby farms or lifestyle blocks where the owners' major source of income is not farming. On average about 10 acres (4 hectares) in size, these holdings are mainly near urban centres. Their prices are related more to the urban economy than to the rural, but when sold, their land prices appear in rural statistics.

Urban expansion can affect the price of farm land if the farmers whose land is taken for such expansion reinvest the money they receive in farming. Suppose, for example, that an airport authority pays two or three dairy farmers big money to obtain their land for an extension to its runway. If they wish to go on farming, these farmers have plenty of cash to buy top dairy farms in the best dairying areas. If they do so, those they buy from also have the money to buy bigger farms elsewhere, and so on. The rural land price stays up, or even increases, but the original money injected into the system has nothing to do with farming.

When farmers are buying farm land, they are less interested in a price per hectare than in the productive capacity of the farm. Dairy farms are often advertised with the price quoted in dollars per kilogram of milkfat produced. Meat and wool farms are often advertised in dollars per stock unit carried. This type of assessment of land values has much more to do with farming and the market.

Even so, the price of rural land in New Zealand, measured in dollars per stock unit for example, has risen at much the same rate as the price when measured in dollars per hectare. Farmers, therefore, have as much to do with keeping land prices up as do other buyers.

Figure 4 (see Appendix) gives the land prices prevailing in New Zealand over the last 24 years. There are some interesting factors to note. The spectacular increase in prices coincided with the growth in subsidies, export incentives and a period when interest rates were lower than the rate of inflation. Prices started to fall two years before subsidies were removed, as buyers realised that all was not well with the industry. The later increase coincided with improved economic management after a period of uneven reforms where farming was more adversely affected than it need have been.

Something that does not really show up in the statistics is the fact that the fluctuation in land prices is not evenly spread across the board. I believe that the best land, which has many use options, never really dropped in price at all. Sales slowed down, but prices stayed up. The poor land, for which there were fewer options and which had been the most heavily subsidised, did at first take a substantial fall in price. However, the improvement in economic management generally has seen an improvement in land prices over all classes of land.

Livestock prices have fluctuated also. They have always fluctuated in relation to market prices for the end product, although in times of subsidy this effect is minimised. With the removal of subsidies there can be greater swings, since the 'padding' has been removed. As mentioned at the beginning of this chapter, prices for some classes of land and livestock did fall temporarily in response to the removal of subsidies, but prices for both land and livestock recovered quite quickly once the whole economy had made the necessary adjustments.

The temporary fall in land and livestock prices did, of course, cause great difficulties for some farmers but, as explained in chapter 9, most farmers were, with some assistance, able to cope during this period. Very few actually left the land.

Averages can be very deceptive when it comes to working out why this was so. On average in 1988, New Zealand farmers had approximately $NZ180,000 of debt. For most farmers, this level of debt is quite manageable, even in a difficult year. This 'average' farmer is, however, quite hard to find. Some farmers are net investors, a considerable number have little or no debt, while about a quarter are very heavily indebted. This means that a difficult period can be weathered relatively easily by the majority, while many of the minority who are carrying big debts face severe hardship. The solution is to deal with the problems faced by the few.

This is what happened in New Zealand. Some people argued that this was favouritism and that many farmers were in trouble because of their own mismanagement. Most, however, considered that the changes being made were extraordinary, that it was purely the luck of the draw as to what stage of your farming life you were in when it happened, and that a once-only adjustment was fair.

A similar situation applies with livestock, where again not all farmers are affected by a drop in prices. A fall in the price of cows, for example, does not affect most dairy farmers greatly. Most of their income is from milk, rather than from livestock sales. A sharemilker selling a large herd in order to buy a farm will be affected, as will a retiring farmer who is selling out, but the majority will be unaffected.

Farmers who are finishing (fattening) livestock can adjust their profit margins if prices fall. They simply pay less for their replacement stock. This leaves the livestock breeder carrying the load. The problem is short-term, and as with land price falls can be dealt with by helping the minority who are affected. The rest can cope perfectly well.

In short, the important points are that the withdrawal of subsidies may have caused a short-term fall in prices of land and livestock, but caused severe problems to few, and they were helped through the transitional phase. The fall was shortlived, and would have been shorter still if reforms had been more evenly applied across the population.

I am confident that a similar transition could be made in other countries, even in those where farmers are heavily subsidised. The scheme put forward by Germany's Professor Tangerman, in which support would be capitalised and paid out to farmers in the form of a one-off grant to see them through the transition, would easily deal with the situation. Farmers could use the money to make whatever individual adjustments were necessary.

This type of solution does imply that farming would become less profitable if subsidies were withdrawn and a temporary drop in land and livestock prices occurred. But this is by no means certain. The New Zealand experience has been that the costs of servicing the farming industry dropped dramatically following the withdrawal of subsidies and the subsequent adjustments.

It is perfectly clear that manufacturers of farm inputs price them at what the market will stand and that while subsidies continue to rise, so will input prices. The reverse is also true. Farm input prices in New Zealand have actually fallen in three of the years between 1989 and 1995. A British friend noted on a recent visit to New Zealand that a container of weedkiller (identical in brand, size and chemical content) was selling in New Zealand for about one-third of the price it sells for in the UK. I believe that such reductions in farm inputs will occur in other countries if subsidies are removed.

It is by no means certain that land prices would fall, either. A well-managed transition to a market agriculture should mean profitable farming, especially on good land once limitations on good farming such as set-aside and quotas were removed.

We have been making the assumption in this discussion that a fall in land or livestock prices is necessarily a bad thing. This is the other great myth which keeps subsidy regimes from being reformed. The chairman of the Australian Wool Board was recently reported criticising government support to farmers in Australia because he believed that it was pushing the price of land above the value of its productive capacity. He said that land prices should be allowed to adjust to lower wool prices, so that wool farming could become profitable again.

A young New Zealand farmer complained to me recently that it was impossible for her and her brothers to take over their parents' farm at present land prices. If young farmers cannot afford to get into farming because land prices are artificially inflated, farms will be sold to existing wealthy landowners or to other wealthy people for uses other than agriculture. This is not the path New Zealand wishes to take.

I have heard farming leaders in other countries express concern for the future of the family farm, and for the young farmers who want to get into the industry. However, their continued support for subsidisation suggests that what they say and what they do are quite different things.

I have already made the point that concessional interest rates for the young farmer were not the solution to this problem, for they increased the price of land and actually hindered the young farmer from purchasing a farm. The only winners were the retiring farmers, who got an undeserved reward when they sold out. This was certainly not the intention of those who brought in the scheme in the first place, and young farmers did not mourn its demise.

For any farmer, the return on capital employed must be of prime importance. When land prices are high, the return on capital will be low for purchasers of land. They may struggle for years to get their heads above water, with much of their income going to service their debts. A more realistic purchase price may not please the vendor, but it will certainly make life easier for the buyer and make the farming venture more profitable.

To sum up, high land prices do not mean profitable farming, and removing farm subsidies will not cause dramatic drops in the price of land or livestock. Temporary and minor falls can cause problems for a few, but these can be dealt with by sensible short-term assistance.

THE MYTH THAT SUBSIDIES CREATE EMPLOYMENT

Wherever you are in the world and whatever subsidy is being promoted or defended, you can be sure that the employment it supposedly provides will be a major reason for promoting it.

Almost every report I see prepared for the British sheep industry by supporters of subsidies claims that there would be virtually no economic life in the hills if subsidies to the sheep farmers were to be taken away. These reports tell of jobs created for farm owners, shepherds and other farm workers, and refer to those who make and sell the chemicals, fertilisers, animal remedies, tractors and other vehicles, tools and other essentials used by farmers. They tell us to take into account those who transport goods to and from farms, the sheep shearers, the butchers, the meat processors, the workers in the woollen mills, and so on. They remind us also about those who teach the children, build and maintain the roads, run the shops, and provide banking facilities.

It is claimed that very few of these people would have employment if subsidies were not paid to sheep farmers. I know these arguments well, because I used them myself in New Zealand, years ago.

Farmers are not the only guilty party when it comes to making such claims. In New Zealand, the motor vehicle assembly industry has been a classic example. New cars in this country used to be among the most expensive in the world. This was because for many years we had a policy of importing only the components and assembling the car in New Zealand. The industry was also given a high level of tariff protection, with the intention of providing jobs for those assembling the cars.

Many battles were fought between those wanting cheaper and better cars and those enjoying the benefits of the protection. Any changes proposed were greeted with howls of protest and wild claims that thousands of jobs would be lost as a consequence. The total number of people employed in the industry was often quoted as the number of possible job losses, in order to deter people from supporting any change.

In fact, the number of people actually assembling cars is a very small proportion of the total industry workforce. There are also all the others who sell, repair and service the vehicles.

If the tariff protection is withdrawn and the price of cars goes down, then consumers have more money to spend on other items. The increase in sales of these goods will, of course, also provide employment.

We have been assuming that the assembling company will not improve its efficiency and compete with imported car prices. If the company in fact does improve efficiency and goes on assembling cars, we will have the best of both worlds. In December 1995, a report in the *New Zealand Herald* proudly stated that the locally assembled Honda Accord cars are being reduced in price. This shows that companies can compete as protection is phased out - they simply need to improve their efficiency.

The principles I have outlined in the New Zealand car industry would also apply, for example, in the sheep farming hills of the UK if subsidies on sheep were withdrawn. The hills would still be farmed. People supplying farmers with goods and services may be forced to reduce their prices, and farmers may also have to ensure that their products are more efficiently produced and better marketed than they are now.

There is no incentive to market your products properly if your income is always assured. The money saved from farm subsidy payments would be available for investment in other areas. Any reduction in the farm labour force would be replaced by job growth in these other areas.

Farmers are also likely to diversify into other money-making ventures. More and more farmers in many countries are, for example, capitalising on the growth of tourism through the setting up of facilities such as bed and breakfast lodges.

Subsidies actually have a very poor record when it comes to protecting rural employment. As mentioned, in Canada and Denmark supply management systems and heavy protection are stoutly defended because they encourage the retention of the family farm - yet farmer numbers have been reduced drastically since supply management was introduced. How their proponents can go on promoting subsidies and protection in the name of preserving family farming, I do not know.

In an article in the *Economist* of 5 October 1991, the Paris correspondent reported that immediately after World War II farmers represented over one-third of the working population in France. By 1960, that had dropped to 20 percent, and in 1991 farmers represented only 6 percent of the workforce. Between one-third and half of the existing farms are expected to disappear by the year 2000. This is in a country where support to farmers is at a very high level.

On the other hand, the predictions of those who claimed that the MacSharry reforms of the CAP would put millions of farmers out of business do not appear to have been fulfilled.

It can certainly be argued that subsidies have hastened the reduction in employment in the farming industry. Figure 7 shows that New Zealand, the only country in the OECD survey that has had subsidies to a high level and has subsequently removed them all, is also the only country where the number of people employed in agriculture (in 1990, as compared with 1977) has not gone down.

Unemployment in New Zealand has been at a level which most New Zealanders would consider unacceptable for the last fifteen years. In 1981 it was measured at 3.5 percent of the workforce, and increased steadily to 5.7 percent in 1984. The period 1985-88 saw it steady at between 4 and 4.3 percent, but it rose to 6.2 percent in 1989, and kept rising until it peaked at 9.1 percent in 1991.

By 1995, the reforms were having a beneficial effect on employment, and the unemployed figure had reduced to 6.1 percent during a period of strong growth. A record number of 148,000 new jobs were developed over the three years 1992-95.

Unemployment is not a problem New Zealand has on its own. It is something most countries are struggling to overcome. In New Zealand we still have some way to go, but if we look at the OECD figures we can see that we are doing comparatively well. Our 9.1 percent to 6.1 percent drop in the last four years or so has lifted us from near the bottom of the OECD employment list to third, behind Japan and the United States.

Furthermore, it is fair to say that the real New Zealand unemployment rate was masked in the early and mid-80s by the employment policies of the large and inefficient public sector. I have been told that it was quite common for Ministers to instruct their departments to hire more staff, even if there was no work for them, to keep the unemployment figures down before elections.

In the late 1970s and early eighties in New Zealand, a scheme called 'Think Big' was central to the National Party policy. It included a number of energy-related projects, and an expansion of the steel mill in South Auckland. The success of the energy projects depended on continued high prices for oil, while the steel mill's viability was based on a high level of border protection.

The success or failure of these projects is still open to debate. On the whole I consider them to have been unsuccessful, but that is not why I am mentioning them. I have no doubt that the number of people employed in their developmental stages kept the percentage of unemployed people temporarily lower than it would otherwise have been in the early eighties.

The reduction in unemployment between 1984 and 1987 coincided with the deregulation of the financial sector, the withdrawal of the wage/price freeze, the 20 percent devaluation of the currency, and a very buoyant international financial sector.

The steep rise in unemployment followed the 1987 collapse of the stock market and the recession which ensued. It also went hand in hand with the long overdue restructuring of the grossly overstaffed government departments, the commercial restructuring of many businesses, the reform of the waterfront industry, and finally a genuine freeing up of the labour market.

Although the initial impact of these moves increased unemployment to the 1991 peak of 9.1 percent, now that business and farming are gaining the benefits of the reforms and the consequent low rate of inflation (0-2 percent), there has been real growth in the job market.

Current Minister of Finance Bill Birch has been a Member of Parliament since 1972. He is credited with being the architect of National's 'Think Big' policies. In a post-Budget speech in Wellington on 1 June 1 1995, he made an interesting remark:

"Today's performance is very different from the short-lived recoveries we have experienced over the past 20 years. It does not depend on extra Government spending, or a big pick-up in overseas prices or demand for a few of our agriculture products. In a major break from the past, the economy is not expected to go into the boom/bust pattern we have so often seen before."

The number of people employed in agriculture as a whole declined by 9,000 between 1981 and 1986. By 1994 it had fallen by another 10,500. This represents a very small reduction (from 20.5 percent of the workforce in 1981 to 18.1 percent in 1994), in 13 years. On the farms themselves, the number of people employed has actually increased by 7,000 since 1981, going up from from 9.2 percent of the total workforce in 1981 to 9.3 percent in 1994.

In 1986, just after the major cuts in support were made, the figure was 8.8 percent of the workforce, and in 1991, when the benefits of all the reforms finally being put into place had not quite impacted on agriculture, the figure was 8.5 percent. To have reached and surpassed the 1981 figure by 1994 is an interesting achievement, and rare in world terms.

The increase in on-farm employment also mirrors a gradual intensification of farming on better land. This includes an increase in dairying and overall horticultural production. These developments are in response to market indications, rather than to government subsidies.

Over the period of change, there were quite spectacular falls in the numbers employed in both the processing and the input supply sectors, as these businesses were forced to reform or go bankrupt. However, employment in these sectors is also now increasing as the more profitable areas of farming expand. Much more downstream processing is also being carried out in New Zealand as this area becomes more cost effective.

The information in this chapter further illustrates that support to farmers does not end up in farmers' pockets. In New Zealand, it was mostly transferred to the inefficient processing and supply sectors. This may have meant more employment in these areas, but it did not create 'real' jobs. We now have real economic growth and real jobs.

The New Zealand experience clearly shows that it is a myth to claim that agricultural subsidies create employment. What they actually create is inefficiency.

It is also a myth to claim that there will be massive job losses in rural areas if subsidies are withdrawn. If the reforms are carried out in the correct way, it is likely that job opportunities will increase over time.

It must be remembered that the period under discussion has been one in which New Zealand also had to deal with markets grossly distorted by protection and subsidisation. If protection and subsidisation schemes maintained jobs in other countries, these jobs have been maintained not only at the expense of their consumers and taxpayers but also at the expense of employment in New Zealand and other similar countries.

Subsidies do not create employment. They simply move resources from one sector to another.

THE MYTHS OF FOOD SECURITY AND FOOD SAFETY

Systems of subsidies and protection were initially developed for farm produce in several countries because governments were persuaded that their people would starve without them. This is the theory of 'food security'.

This fear of starvation stems from the great wars of the first half of this century, when hunger was common among people on both sides of the conflict. A number of the countries concerned were partly dependent on food imports, and when ships were bombed or mined supplies were disrupted.

It was decided that such a situation should never be allowed to happen again, and the food security policies were developed. These systems may have had some validity at the time, but to suggest that the same grounds apply today is certainly a myth. In fact there would be greater security today if there were more food trade, rather than less.

Another myth concerns food safety. This would have people believe that food produced locally is safer, in this health-conscious age, than imported food.

I believe that what has become known as the system of comparative advantage provides much greater food security for the people of today's world than does the system of protecting local farmers from competition. Comparative advantage is practised in New Zealand. We grow products with which we can be internationally competitive. We import products that farmers in other countries can produce more efficiently than we can, and for which there is a demand.

Tropical fruits are a good example. Either we cannot produce them at all, or we can produce them only at great expense. For other products, the decision is not so clear-cut. We could grow sugar beet as cheaply as it is grown in some other countries. We cannot grow it cheaply enough, however, to justify establishing it as a major industry in New Zealand, although several campaigns have been mounted to do this. I am pleased to say, however, that common sense has prevailed. The fact is that many countries can produce cane sugar more cheaply than we can grow beet sugar, so we import our sugar.

Sugar is only one example of New Zealand's 'open door' attitude to non-subsidised food imports. If we expect to export our own non-subsidised products to the world, it would be totally untenable for us to try to restrict imports of similarly unsubsidised food.

There are some who would go as far as to say that we should open our borders to subsidised products as well. They say that if other countries are stupid enough to subsidise their producers and exporters, then we should take advantage of the cheap goods. However, farmers have drawn the line at this one.

The system of comparative advantage has real benefits to consumers, as it means that they can buy quality products at cheaper prices than they can under a system of protection. It is also good for farmers, who can concentrate on producing what their climate, soils and skills suggest they should be producing, and still make a profit.

In such a system, market prices will dictate how much is produced, whereas in a protected and/or subsidised environment it is likely that politicians and bureaucrats will decide. They may also decide what prices will be paid, and could set them at a level unsatisfactory to farmers.

War is a terrible thing, and it would be good if there were no more wars. However, the fact that a country produces its own food is no guarantee that it will have adequate food supplies in times of conflict. Imagine what germ warfare could do to a nation's livestock. Think about what nuclear fallout could do to crops and flocks or herds. Imagine what damage hormone sprays could do to food production, if they were used as a weapon. Chernobyl proved that we do not even need war to have nuclear fallout. Any major nuclear accident could destroy a nation's food supply in moments.

Should such a disaster occur, countries would be much more secure if they had been trading with others, and knew that these others could respond quickly with increased food trade where necessary. A country with established trading patterns would have little difficulty in getting such a response, whereas a country which has always kept out food imports would have a huge problem on its hands.

Of course, something as rare as a war or a nuclear accident is not necessary to cause food supply problems. The weather can have devastating effects also, as can disease in animals. Look at what 'mad cow' disease is doing to the British beef industry in 1996.

Japan, a small mountainous country with a large population, has become a large importer of food. It imports approximately 50 percent of its food requirements at present. For its own reasons, however, it has gone to great lengths to be self-sufficient in rice. Japanese farmers have been paid up to seven times the world price of rice in order to encourage them to produce enough for their country's needs.

Poor growing weather in the 1994 season produced a low yield and left Japan short of rice. For the first time for many years, they had to go to the international market to feed their people. A real problem ensued because the only rice available was the longer-grained variety preferred in other countries. Japan had made such an effort to keep out imports of short-grained rice that no-one else bothered to grow it much of it. The Japanese, therefore, in a time of shortage, had to use a type of rice they did not like, or go hungry.

Had Japan been prepared to import some short-grained rice every year, there would have been growers around the world who could have supplied it during shortages, even if they had to store it from time to time. Consumers in Japan would always have what they prefer, and farmers in other countries would also be better off. The only losers would be the Japanese rice growers, whose prices would fall, but as they are among the wealthiest farmers in the world, there is some room for adjustment.

Contrast this with the sheep-meat situation in Japan. Overall the Japanese do not eat much sheep meat, but on the northern island of Hokkaido there has always been a tradition of sheep-meat eating. Those who have eaten Genghis Khan-style in the beer gardens of Sapporo will know about this tradition.

The inhabitants of Hokkaido almost starved during World War II, and they slaughtered all of their sheep to keep themselves alive. The sheep have never been replaced, but the residents of Hokkaido have never since been short of sheep meat because producers from around the world compete strongly to supply the market. There will never be a shortage of products for which people are prepared to pay.

This means that food will be produced in countries where subsidies and protection are now used, even if the system is changed and these mechanisms are phased out.

It is absurd to think that there would be no farming in the San Joaquin Valley in California or the Paris Basin in France if there were no subsidy support. In fact, these areas would probably produce more, because there would be no restrictions such as quotas and set-aside to hinder production.

Countries such as Norway, Sweden and Finland would not stop producing food because imports were allowed. There are many consumers in those countries who, for a number of reasons, prefer local food, irrespective of cost. Imports are likely to increase, however, as other consumers respond to the price and the variety offered.

On the other hand, exports will also increase as farmers are forced to examine what they can produce competitively, and begin to adapt their farming methods accordingly. Some of the best strawberries I have ever tasted are grown under the midnight sun in Norway. This part of the world can produce berry fruits when they are past their best in the rest of Europe, and out of season in most of the rest of the world - a fact that strongly suggests there must be an international demand for them.

Reindeer farming might also be a growth industry if the dairy, sheep and grain farming alternatives were not so heavily subsidised and protected.

These are just a few examples of many to show that more liberal trade would improve food security, and provide more variety for the consumer at more reasonable prices. The rewards to the farmers would also be greater overall, as they would be producing to their comparative advantage.

As a New Zealand farmer, I have been on the receiving end of problems caused by the protected, subsidised farming regimes practised in the wealthy industrialised countries under the guise of food security. We have suffered as access for our produce has been cut, and we have had to compete against exports of artificially created and dumped surpluses.

Substantial as these problems have been, they pale by comparison with the devastation caused by the subsidised, protected sugar regimes of North America and Europe in the sugar-dependent tropical countries, such as the Philippines, Fiji and countries of the Caribbean. Without warning, these countries were faced with an almost total cut in demand for cane sugar as it was shut out of the market, with consequent bankruptcies and huge rises in unemployment.

On a small island in the Philippines, a group of sugar growers actually starved to death before the authorities could give aid.

To add insult to injury, American and European organisations then sent aid to the countries their subsidised beet sugar and corn syrup regimes had ruined. The aid amounted to about 1 percent of the money that the cane sugar had sold for in those countries.

A small share of the sugar market has been provided for some of these countries in subsequent international sugar agreements, but had the comparative advantage system been allowed to operate throughout, everyone would have been vastly better off.

I object very strongly to the claim that imported food is not as safe as that grown locally. The EU and the United States both have strict safety standards with which all countries that export to them must comply. We also have Codex Alimentarius, the international safety standards that have been used as the basis of the GATT sanitary and phytosanitary agreement.

There has been a real need for standardisation through the GATT, as some countries are tempted to take regulations to extremes and use them as trade barriers rather than safety standards.

Spreading myths and rumours is the hallmark of people opposed to more liberal trade. I was very annoyed at several meetings in Europe in 1990 to hear an anti-GATT campaigner tell audiences that Codex was run by multinational chemical companies that created the rules for their own ends. This was a blatant untruth told to encourage people to suspect imports and resist change. Such tactics are especially galling to New Zealanders, who have complied with regulations from every other country for many years and at huge cost.

Lying by omission is another tactic. I attended the National Farmers Union of England and Wales conference in 1992, and heard a delegate complain that local abattoirs were having to comply with what he considered to be unreasonable EC meat hygiene regulations. Hundreds of abattoirs throughout Europe were operating at standards well below what the regulations demanded. Many millions of pounds were required to upgrade or replace those in the UK alone. The punchline from the delegate was that he 'bet that the New Zealanders did not have to comply with the regulations!' The NFU leaders sitting at the top table were all well aware that New Zealand had complied with all regulations for over 20 years, but to a man they sat in silence and let the comment be accepted as the truth.

I was able to put the delegate and his colleagues right on that occasion, but had I not been present many conference participants would have gone home ignorant of the facts.

The fact that New Zealand had been complying with the regulations for years, while much of Europe did not, was really brought home to British sheep farmers when a French importer told the centennial conference of the National Sheep Association in 1992 that New Zealand abattoirs were 'light years ahead' of most in Europe, which was why the product was better. This is not an isolated case. In many countries imported food is safer than that produced locally because it has to comply with more stringent regulations than those imposed on the locals.

Many exporters are now moving beyond simply complying with safety regulations and seeking ISO quality accreditation. They are conscious of what consumers are demanding and want them to know that their products are not only safe, but are also of the highest possible quality.

A group with vested interests has gone to great lengths to perpetuate the myth that more liberal trade will reduce food security and food safety, when in fact the opposite is true. More liberal trade will improve world food security and lift standards of food safety. Vested interests have no place in an argument as important to people's health and peace of mind as this one.

THE MYTH OF THE MULTINATIONAL TAKEOVER

The idea that the multinationals will take over farming in any country that gets rid of subsidies is another myth spread by those opposed to subsidy reduction and the development of more liberal trade in agriculture.

Leaders of the American Farmers Union have strongly supported the belief that the only the multinational commodity traders would gain from subsidy reduction, and have promoted a fear that family farmers would be at their mercy. Opponents of change also argued that the multinational companies favoured reform because it would enable them to 'blast open' (their term, not mine) the markets of the Third World, take over the land and force peasants to work for them, rather than grow their own crops.

Multinational companies are already involved in developing countries, dealing in agricultural products such as bananas in Ecuador, oranges in Brazil, and pineapples in the Philippines. This seems to have happened with the approval of the governments concerned and may have improved the productivity of local industries. Whether or not this involvement is beneficial to the farmers of the developing countries, it has taken place before trade liberalisation has even started. It is a moot point whether more or less of it will happen with subsidy removal and greater market access.

The argument that multinational companies would take over the Third World economies under trade liberalisation is surprising in the light of the Cairns Group's approach to the GATT round. Most of the Cairns Group countries are in the developing category, and their leaders believe that it is in the best interests of their people to support liberalisation. They want more access to world markets so that they can export and improve their farming incomes. When they see the wealth that industrialised countries have built up through trade, there is bitter resentment when markets are closed to their agricultural products. They certainly cannot afford to travel the subsidy route.

'Blasting open' the Third World markets seems a rather strange way of putting it. Many of the markets of the developing world are already more 'open' than are the protected wealthy ones of Europe, Scandinavia and Japan. Even if big companies were able to get better access to these Third World markets than they have now, how many people would be able to afford their products?

To talk of fearing multinational takeovers of land in the Third World perhaps makes more sense. People might fear the loss of their land, or the use of their low production costs to export back into the wealthy markets of the world, without giving the local people much in return. Farmers in subsidised countries may fear the competition of such moves. While liberalisation and subsidy reduction are still in their infancy, it is probably too soon to be making final judgements on this aspect. We can, however, make a judgement on the case of New Zealand, which is the one country to have had a subsidy system and then to have dismantled it completely.

It is now eleven years since the first subsidies were removed in New Zealand. If removal of subsidies was going to produce a multinational takeover, then surely agriculture in a country as small and vulnerable as New Zealand would have been completely taken over years ago.

Although some of the former government-owned businesses such as Telecom and NZ Rail have been sold to international companies, there is very little of what could truly be called multinational involvement in the farming sector. Dairy processing and marketing is totally New Zealand-owned, as are most dairy farms, although a handful of farms have overseas owners. The marketing and most of the processing sectors of the kiwifruit and apple industries are New Zealand-owned also.

The Dairy, Kiwifruit and Apple and Pear Marketing Boards are protected by statute, so it would not be possible for multinationals to become involved as outright owners in many areas of these industries. The marketing boards do have dealings and joint ventures with them, however. The Kiwifruit Marketing Board's Japanese marketing operation, for example, was handled by Dole Japan in 1995.

The single-seller aspect of these marketing boards is somewhat controversial in New Zealand, although not with the majority of farmers.

Most dairy, kiwifruit and apple producers strongly support single-selling, and want it to continue. Some meat and wool farmers would like to see a similar system in place to market their produce but they do not appear to be in the majority at the time of writing.

Criticism of the boards and single-selling does not come from the multinational companies, either. It comes from the ideological economists, the self-styled entrepreneurs who would like the chance to take the easiest pickings from the farming sector, and from a few farmers who would also like to get into the wealthiest markets on their own account.

There are, by contrast, no wool or meat marketing organisations protected by law from overseas ownership or internal competition. Entry is completely open. There has been foreign ownership of meat processing and marketing companies for many years. At one time overseas owners probably predominated. In recent years, however, this has changed. The last major overseas-owned meat company, Weddel, which was part of the Vestey organisation, went into receivership in 1994.

This does not mean that overseas investment is not welcome. In fact, overseas companies that can assist in the development of markets for our meat are welcomed with open arms. In a number of the wool exporting and marketing companies, for example, the overseas owners play an important role in marketing the product.

It is interesting that since subsidy removal, the trend has been strongly towards a return to New Zealand ownership in the meat industry in particular, rather than more overseas ownership as the myth-makers would claim.

Watties, a major New Zealand food processing and marketing company, was taken over by Heinz a few years ago. This involvement was welcomed both by Watties and the farmers who supply the company with its raw materials. Interestingly, as previously mentioned, Watties-Heinz has made more progress in recent times than any other operator in New Zealand in organic farming, which is often seen only as a place for the small farmer and the niche marketer and processor. The organic venture has been taken on with the enthusiastic support of the growers concerned.

A number of international chemical companies operate in New Zealand, as they do in many other countries. They have not noticeably increased their market share since subsidy removal. In fact, some local farm-associated companies have expanded overseas. In a bold move Fernz, a fertiliser and chemical manufacturer, has expanded to Australia and beyond. The farm service sector is mainly New Zealand-owned.

There is some corporate farming in New Zealand. One of the corporates, Tasman Agriculture, has the New Zealand listed company, Brierley, as its major shareholder, but also has overseas shareholders. It has bought farms in Australia, but can hardly be considered a multinational yet. Brierley itself has a wide variety of investments in a number of countries. Its New Zealand involvement includes some agriculturally-related investments, in addition to Tasman Agriculture.

Corporate farming has always been somewhat controversial, but it appears to be gaining in acceptance, although it is still small by comparison with family farming. I believe it is more acceptable now because it is seen as providing excellent job opportunities for ambitious young farmers, especially in the dairy sector.

The biggest farming enterprise in New Zealand is the government-owned corporation Landcorp. It is what is left of the large parcel of land that the government developed, under the previous policies, to settle young qualified farmers on farms. There is talk of breaking up the organisation and selling the land but there is no strong pressure for this to happen. The corporation is able to use its large pool of genetic material to produce superior animals and it is feared that this resource would be lost if the organisation was sold. There is, therefore, some support for retaining it.

Some of the strongest corporate farms are owned by the Maori people. Their system of multi-ownership is well suited to the corporate system, and some of their corporations are being developed to the point where they can control the whole operation from farm gate to marketplace. This development has many positive aspects, and money made from the corporation is often ploughed back into education for their young people.

In forestry there is both corporate and multinational ownership. One of the two companies dominating the industry, Carter Holt Harvey, has majority American shareholding while the other, Fletcher Challenge, is New Zealand-controlled but has some overseas shareholding.

There is very little controversy over such corporate or foreign ownership. The investment required to harvest and process the very large forestry crop is beyond the resources of small operators. It may even be beyond the resources of the New Zealand Government, because it is considering selling its forestry interests. But this does not mean that individuals cannot participate in the forestry industry on their own account, or through a great variety of partnerships and syndicates.

Most of the banking industry and many of the insurance companies in New Zealand are overseas-owned, but this has always been the case.

The main change in the past ten years has been the government's sale of its own banks, finance and insurance companies. These have largely gone to off-shore interests. The Rural Bank was initially sold to Fletcher Challenge, but it has been sold on to the UK-owned National Bank of New Zealand. The sale of the Rural Bank to Fletcher Challenge, ahead of a farmer bid (led by Peter Elworthy and Ken Macdonald) caused considerable upset at the time, but the anger has since largely dissipated.

There is strong competition among banks and stock firms for farmer business. Trust Bank, a growing New Zealand-owned bank, is increasing its rural involvement. The stock and station firms are all New Zealand-owned now, where some years ago there was considerable overseas ownership. A strong farmer-owned co-operative insurance company has a large share of the rural insurance market.

Most farms are still held by farming families. The commercial farms have become larger as technological developments are applied. Farming is much more mechanised than it once was. Machinery is more sophisticated, and computer technology is commonplace. It is a totally different industry from the one my father worked in. Much of the drudgery has gone. Cows are easier to milk, hay and silage are harvested totally by machine with very little physical effort. Management techniques have been developed to the point where our farm adviser tells us that we need not make hay or silage any more, and so on. It is therefore hardly surprising that farms are getting larger, and one farmer can run more and more stock alone.

Except for forestry, the larger farms are not the result of corporate buying. They are simply the result of competent farming families expanding their operations. At the same time there are still people running small properties in a multitude of ways. Despite all the modern systems that are available, some people like to do things the old way. There are still a few milking cows and goats by hand, and good luck to them. If they wish to live and farm that way, they are perfectly free to do so.

It could perhaps be said that the removal of subsidies from sheep farming has made the expansion of the forestry industry easier than it would otherwise have been. However, most of the land in question is better suited to growing trees than it is to grassland farming. There will probably always be debate between farming and forestry interests over the use of areas.

In summary, it can fairly be said that the ownership of New Zealand's farms and farm service sectors is diverse. It is a huge assumption to lump all multinational companies together in the 'undesirable' category.

Many of them make a positive contribution to farming.

The strengthening of the New Zealand economy has meant that some farming companies and individuals have expanded their operations into other countries with considerable success. Perhaps some of these will be the multinationals of the future.

If, however, a fear of multinationals is holding back any government from liberalising its farming industry, the New Zealand experience clearly shows that this fear is unfounded. There will be multinational activity where the companies consider it appropriate, with or without subsidies and protection. Most multinationals do extremely well with subsidies and protection. They may or may not do so well without them.

I often think back to the comment made by the chairman of the Rabo Bank when I was seeking his support in the battle for trade liberalisation during the Uruguay Round. He agreed that they should support our campaign, but warned me not to expect any great enthusiasm for it.

'Be assured,' he said, 'we will continue to make money whatever trade system prevails.'

THE MYTH THAT EVERYONE OWES FARMERS A LIVING

What is it that makes farmers in many countries think that they deserve a good income, whether they are efficient and competitive or not?

What is it that makes farmers in the world's wealthiest countries accept that they should be subsidised and protected as of right, while other sectors of their economies are expected to compete on international markets?

What is it that makes dairy and grain farmers in the world's richest and most competitive country believe that they need very substantial export subsidies in order to compete with farmers from other, less fortunate countries?

Why is it that kiwifruit growers in the United States want their government to apply punitive tariffs against New Zealand fruit because it was once sold for a lower price in the United States than it was in Japan? I am sure that United States exporters sell products at different rates in different markets, since it is standard business practice. Obviously it is a case of 'don't do as we do; do as we say'.

How is it that consumers, industrialists and taxpayers in the wealthy countries of the world allow transfers of huge amounts of money from their sectors to the farming sector, with barely a whimper of protest?

Why do the people in disadvantaged countries not fight these and other outrages with more passion and vigour?

Why are people in places like Brussels so passive when demonstrating farmers trash their cities and destroy their property?

What makes the citizens of France stand idly by while farmers burn imported products in the streets ?

The myth that society owes farming a living irrespective of the level of efficiency has a lot to answer for, and is not confined to the wealthy countries. There is growing evidence of similar attitudes arising in newly industrialised countries.

In Korea, for example, a Minister of Agriculture was pelted with cow dung at an angry farmers' demonstration because he suggested that there might be changes to the agriculture regime. The average herd size in Korea is about three cows, which means that many farmers have only one or two. Cropping and rice farms are also tiny. The farmers have seen the growing affluence of the city dwellers and they want equivalent pay rises. This may be understandable, but how can a country afford to pay the average city wage to every person with one or two cows ?

Farmers all over the world have a reputation for being happy only when they are complaining about something. We are not immune from this peculiar behaviour in New Zealand. For example, in a recent letter to the editor of the *New Zealand Farmer*, a correspondent claimed that city people were generally paid salaries that farmers could only dream of, and this was most unfair as farmers had hundreds of thousands of dollars tied up in their farms while city people had no risks at all. Until farmers received a return on their capital and a generous reward for their labour, he would continue to complain.

This is not an uncommon view. Farming organisations in many countries have policies based on the notion that farmers 'deserve' a good return on capital and a fair reward for their work. In such a country it would seem very bizarre indeed if those involved in any business outside farming went to the government for assistance because they believed they deserved it - never mind that their business was inefficient and uncompetitive.

This view that agriculture is owed a living might once have prevailed in New Zealand also, and still rears its head in letters to farming newspapers, but it would be hard to sustain now. Farmers in our country have, by buying and selling, put up the price of land by 100 percent in the last five years. While our complaining correspondent feels that he can only dream about making a good living, there are obviously thousands of others who are prepared to pay the money and give it a go.

The attitude that 'our country can afford to pay farming subsidies, so why shouldn't we take them' prevails in many countries. What is it that makes farmers consider that a good income is theirs of right - even at the expense of much poorer consumers and taxpayers? What is it that makes politicians in so many countries accede to their demands?

There are several reasons. Food is, of course, a basic necessity, so those producing it do perform a vital function in society. The environment is becoming increasingly important, and farmers claim more and more often that they are the guardians of the environment. In countries like New Zealand, farming was for many years almost the only provider of overseas funds, and it thus had a very important role to play.

In many countries, the farming vote has been vital to political parties wishing to win elections or stay in power. Even in countries where farmers are in the minority, their vote can be crucial. The situation is exaggerated in countries such as Japan, where a gerrymandered political system means that rural votes have much more value than urban ones.

Farmers in many countries have done a good job of persuading their governments to give in to their demands for protectionist policies. Perhaps politicians have been driven by the fear that family farming would collapse and there would be an avalanche of rural people moving into overcrowded cities and asking for social welfare and unemployment payments.

Whatever the reasons, the protectionist policies have failed. Much of the money has been captured by the large landholders, and a considerable amount has disappeared into the hands of the fraudsters. Huge and costly bureaucracies have been created, farmers in poorer countries have suffered, and, through it all, the drift to the cities has continued unabated.

There is no doubt that all the subsidisation and protection has made a few people very wealthy, but what actual benefits has it brought to world agriculture? It has certainly brought a vast government involvement to the business of farming in many countries. It has caused distortions in the marketplace. It has meant that consumers have paid more than they should have for many food items, and in some instances they have paid too little. This has in turn led to poor town/country relationships. It has caused food mountains and wastage in some countries and famines in others.

In effect, the myth that everyone owes agriculture a living has produced a farming industry that tries to 'farm' governments, but ends up being taken over by governments. The subsidies have become the market, instead of the consumers. This has left farmers out of touch with the real market, and very vulnerable to political change and to competition from alternative products.

The edible fats market in the UK is a perfect example of vulnerability to competition from alternative products. Over the last twenty years, butter's share of this market has dropped drastically. New Zealand's butter access has been reduced by 50 percent over this period, yet the New Zealand market share has remained constant.

The fact that New Zealand butter has been able to retain its market share seems to have been of greater concern to some British dairy farming leaders than the fact that butter has been losing out to alternatives. They appeared to think that if they could get New Zealand butter out of the marketplace, their own butter sales would increase. Indeed, it is likely that if they had succeeded in getting rid of New Zealand butter, their own market share would have dropped still further, as the New Zealand Dairy Board, through its Anchor brand, is in my view by far the strongest promoter of butter in the UK market.

The attitude in many protected markets is that the consumer can afford to pay more, so the answer is to keep the competition out and the price up. This totally ignores the fact that with many products, and butter is one of them, consumers have alternatives. They can eat other spreads, thus cutting out animal fats altogether.

Many European lamb producers follow a similar line. They argue that if they can keep out imports, especially of quality chilled lamb, they will be able to raise the price for their own lamb and improve their incomes. Such an approach is a one-way ticket to oblivion. It totally ignores the efficiency and marketing ability of the producers of alternative meats. It will certainly lead to less buying of the more expensive lamb cuts, and a loss of market share.

Dependence on subsidies and protection puts a farming industry at a great disadvantage when negotiating charges for goods and services. If you are neither internationally competitive nor particularly efficient, how can you expect your service sector to be efficient?

People with a product to sell do not just charge the customer the cost of production plus a margin for profit. They charge the customer what the market will stand. Compare the different prices umbrella vendors in major cities charge for their goods on wet and on fine days, and you will see what I mean.

Amazingly enough, many farmers who are the beneficiaries of subsidies are the first to complain after returning home from a trip to New Zealand, where they have seen goods priced at a much lower level than they are used to paying.

Farmers are very dependent on an efficient and competent work force to get their products to consumers in good shape and at a competitive cost, if their products are to sell well. How can farmers who are beneficiaries of subsidies criticise trade unions for demanding more pay with no increase in productivity? Sadly, of course, they do.

Such hypocrisy just doesn't wash. We have been there and proved that. When our sheep farmers were receiving subsidies, industrial relations in the meat industry were atrocious. Two-thirds of the total work days lost in the whole economy came from the meat processing sector. Each time the farmers got more money from the government, the work force in the processing industry took industrial action to obtain what they considered to be their fair share of the increase. The price to farmers was guaranteed, so the owners of the processing plants could increase their charges and recover any losses. The whole situation became a vicious circle, and resulted in complete chaos.

We now have a situation where farmers, processing plant owners and the people in the work force are all aware that if they do not get a good product to the market at an affordable price, there will be no farming businesses, processing plants or jobs. Industrial relations are much better, and the industry is much more stable.

Sometimes you just can't win. I was quite proud that we had made our farming business efficient without subsidies, and was flabbergasted to be told by a Swedish farmer that because we had low costs in New Zealand, we were now unfair competitors! This is like saying that because Third World countries pay low wages, they are unfair competitors in markets where much higher wages are paid. Actually, they pay low wages because they cannot afford to pay more, and as long as they are kept out of wealthy markets they will not be able to afford higher wages for their workers.

The sooner farmers everywhere can cast off the mentality that everyone owes them a living, and concentrate on satisfying their consumers and competing in world markets with products they are producing efficiently, the sooner sanity will be restored to the industry.

THE MYTH THAT SUBSIDIES HELP FEED THE WORLD

'I thought that they wanted food.' This was the comment of Pat Shaw, a British farming friend as he faced the prospect of his wheat harvest going to join the huge grain mountain, and quotas on subsequent crops. He made it when millions of people were facing starvation in Ethiopia. The paradox of over-production in half of the world, while millions go hungry to bed every night in the other half, continues to worry many of us, especially those who produce food.

The purpose of this chapter is not to try to solve the problem of food for all the people of the world. I wish that I had a simple answer to that. Rather, it is to explore the issue of whether the world hunger problem would get better or get worse if farm subsidies were withdrawn in the wealthy countries.

We must look at this problem seriously as world population is expected to double to approximately 8 billion part way through the next century, before it levels off. If we cannot feed the current population adequately, how can we feed double the number of people in fifty years' time?

Many people have argued, and still argue, that continued subsidisation and increased food aid will be necessary. I grew more and more concerned at the number of churches and aid organisations taking this stance during the GATT negotiations. It became very difficult for me, as a regular churchgoer and a dedicated advocate of subsidy removal and trade liberalisation, to hear my own ideas attacked from the pulpit.

It is certainly the role of the churches to defend the poor and to try to improve their position in society. It was, however, especially disconcerting to see so many church leaders taking such a strong anti-GATT stand, when I believed that in doing so they were actually working against the best interests of the poor.

I decided that I should meet as many of the people concerned as I could, and talk the issues through. It was soon very clear that the anti-GATT lobby had done an excellent job of getting churches and aid organisations to accept their point of view. The people in this lobby group naturally enough came from sectors that held privileged positions under the existing regime, but they much preferred to have church leaders and people from the aid organisations pushing their point of view. This meant that their selfish interests were given a veneer of morality and concern for those less fortunate.

Scaremongering was the principal weapon in their arsenal, and its use was rife. Typical of the arguments propounded was the notion that the former communist countries would flood Europe with cheap food, and the Western European countries would no longer be able to afford to spend adequate amounts on aid.

Similar things were said about what the South American countries would do to North America under trade liberalisation, and it was claimed that the USA and Canada would not be able to help the Third World with aid projects.

I do not have first-hand experience of working in countries dependent on aid but I have listened to, and been impressed by, many who have done so. There is absolutely no question that what has been done for these countries until now has not solved the problem.

Famine has been temporarily relieved in a few places from time to time. Special fund-raising efforts such as Band Aid by Bob Geldof in the eighties are remembered, but are not permanent solutions. Food aid has filled some hungry mouths for a short while, and aid organisations have worked very hard to get food donated and distributed. By and large, however, the world has failed miserably in solving the problem of world hunger.

What then is the long-term solution? I will never forget the words of Kenneth Kaunda, then President of Zambia, when he told the delegates at the World Food Conference in 1988 that the developing countries needed 'food trade, not food aid'.

Her Royal Highness the Princess Anne, who has worked tirelessly for the Save the Children Fund, made a similar point when she was visiting New Zealand in 1989. Speaking about markets for products from the Third World, she said, 'If you don't have access to markets, you don't stand a chance.' When I met the Princess in London, and explained what I and the Cairns Group were trying to do, she concluded our interview by saying that she could only hope that we were successful.

It seemed to me that the situation existing at the time, where the rich countries denied the poor countries access to markets or at best gave them token access and then offered them food aid, was the ultimate in patronising insults.

In product areas where developing countries could be competitive, they had been further squeezed by the subsidies paid in developed countries. The sugar issue, which I have written about in a previous chapter, is probably the worst example. It is no exaggeration to describe this whole process as an international outrage.

The Cairns Group of countries, which ended up being the strongest force for trade liberalisation in the GATT round, comprised mostly developing countries. Their leaders certainly believed that improved market access was the most important factor for the development of their countries. Food aid provided from subsidised surplus production was definitely not what they wanted. A fair chance to trade with the wealthy countries was their only hope of making progress.

There are many problems with gifted or subsidised food for poor countries. The first is the moral one, where the policies of the wealthy country giving the aid have probably caused at least some of the problems for the poor country in the first place. Possibly it is expedient for wealthy countries to keep other nations poor and dependent, but it is certainly not ethical.

The second is that much more food is required in poor countries than can be supplied by food aid, even with subsidies. What these countries require is the development of their own viable farming systems, in which local farmers are paid a reasonable price for their produce. This will not happen while a country receives food aid, because local farmers cannot compete with free or heavily subsidised food. If they do not receive enough for their crops one year, they cannot afford to plant for the next, and are therefore dependent on food aid again. And so the cycle goes on.

A third problem is the amount of fraud and graft involved in the importing and distribution of food aid. In the Melbourne newspaper *The Age*, on 23 December 1995, Padraic McGuiness wrote: 'the net effect of much international aid has been accurately described as taking money from poor people in rich countries and giving it to rich people in poor countries.' This sounds like a brutal assessment but it is basically a fair one.

When my colleagues and I talked with church leaders and people from aid organisations about the failure of food aid and subsidised surpluses from wealthy countries to solve world hunger, most of them were prepared to discuss our philosophy. A number saw what we were trying to do, and applauded it, while others acknowledged that there might be two sides to the issue.

Another group, however, was not so ready to listen and I came reluctantly to the conclusion that some aid workers did not want to see the Third World in anything other than a dependent position. They seemed to enjoy dishing out aid and appeared to have a vested interest in keeping things just the way they were. This may seem a harsh judgement, but one aid worker actually said to me, 'But if we did things your way, I'd be out of a job.'!

I do not claim for a moment that developing farming in Third World countries will be easy, nor do I lay all the blame for world hunger on the wealthy countries. Much of the suffering and devastation has been caused by warfare within the poor countries. However, apportioning the blame will not improve the situation. We have to get on with solving the problem.

Developing farming is not as simple as paying local farmers better prices for their produce. Whole infrastructures must be set up. Capital must be found to water and fertilise the land, to plant crops and shelter, and to fight disease. It is an immense task.

However, as the saying goes: 'Give a man a fish and you will feed him for a day. Teach him to fish and you will feed him for a lifetime.' The principle applies equally to farming. Give a family a food parcel and you will feed them for a day. Teach them how to farm and they can feed themselves for life.

I have barely scratched the surface of this huge issue. Other big factors come into it, such as population control, which some people advance as providing the whole solution. Even if this were so in the long term, it would not deal with the problems of the next half century.

What is clear, though, is that the problem of world hunger has not been solved under subsidies and protection. Indeed, it can be argued strongly that they have actually hindered progress, in which case subsidy removal and liberalisation of trade can only improve the situation.

Under a more liberal system, farmers in developed countries would be producing for actual markets, not for government intervention schemes. There would be far fewer surplus products to distribute, and far more government money available for aid, amongst other things.

Local farmers in poor countries could get reasonable prices for their crops, and aid (in the form of money and expertise, rather than food) could be used to establish productive farming throughout the Third World.

In the short term, emergency food aid would still be needed in specific areas. This could be supplied by a world food bank, set up for the purpose, and distributed carefully so as not to ruin local farmers.

Let us not fool ourselves. Trade liberalisation and subsidy removal do mean a reallocation of the world's resources. The great problem has always been that a large percentage of these resources have been held by a favoured few

The World Bank Atlas estimates of average national income for 1994 have recently been published. These figures show a rise in average income worldwide of US$60 per year. However, the poorest country in the list, Ethiopia, has an average income of only US$175, and incomes there have declined every year since 1985.

Contrast this with the wealthiest countries on the list - the top six are Luxembourg, Switzerland, Japan, Denmark, Norway and the United States - with average incomes from US$53,700 to US$34,850 - and we can see that citizens of the rich countries have incomes of 200 to 300 times those of their Ethiopian counterparts.

An interesting footnote to the reporting of these figures in an Australian newspaper is the priority we apparently place on items of news. The piece about Ethiopia's increasing poverty rated only three lines of print. The controversy about the proposed divorce and possible remarriage of the Prince of Wales rated three whole pages.

The few favoured and wealthy countries that control the world's resources have used agricultural protectionism to help retain their position. It is hardly surprising that many people from these countries opposed the GATT proposals so vigorously. Nor is it surprising that some of their citizens are so vociferous in their opposition to the removal of subsidies.

Their attempts to convince the rest of us that they oppose these things in the name of the poor people of the world, however, should no longer impress anyone. Protectionism and subsidisation are designed to make the rich richer and the poor poorer. That is the truth, and no amount of hypocritical 'concern' for the poor of the world on the part of advocates of these things can ever disguise it.

The myth that subsidies help feed the world is probably the most dishonest and damaging myth of all.

FARMING WITHOUT SUBSIDIES - A DECADE ON

Those involved in the reform process in New Zealand are justifiably proud of what has been achieved. New Zealand has gone from having one of the worst records of OECD countries in key areas such as inflation, productivity, competitiveness and employment, to being one of the better performers. We are achieving budget surpluses at a time when countries with much greater resources are still in deficit. Our farming industry is productive and viable without subsidies.

Both Moody's and Standard and Poors upgraded New Zealand's credit rating early in 1996. This has put our rating above that of Australia (a country with much greater natural resources) for the first time ever. Anyone who understands the relationship between Australia and New Zealand will appreciate how galling this would be to an Australian commentator. Terry McCrann, writing on the business page of the Melbourne *Herald Sun* on 5 January 1996, says:

"Let that sink in: poor, backward, country cousin New Zealand regarded as more creditworthy than Australia. . . . Especially as, less than a decade ago, NZ was the sick, sick country of this region - seemingly destined to slide rapidly into serious national poverty and consequential serious social conflict. Yet precisely because NZ has embraced the reforms necessary to make it a successful player in the global marketplace - the reforms which the Keating government has placed in the too-hard basket - it is now building a sustainably prosperous future."

Visitors come to our shores to see how it is done. The World Bank, for example, has brought decision-makers from many countries to study the reforms. There is a continual procession of overseas television crews who make programmes on farming without subsidies. A number of New Zealanders have been invited to speak on the subject in other countries.

One thing that both the visitors who are reporting back and New Zealanders who are lecturing overseas have to be careful about is giving the impression that it all came easily. It certainly did not. In writing this book my intention has been to concentrate on the farming reforms. Giving a detailed account of the changes that took place in many other sectors would occupy several more volumes, and so this is not the place to do that. We must recognise, however, that most New Zealanders have faced change in one form or other and that there has been genuine hardship. As I have pointed out in earlier chapters I believe that the situation for some, especially for farmers, was more difficult that it need have been.

There have, naturally enough, been some strong critics right through the process. I do not go along with those, and there are a number of them, who label the reforms as an experiment. Several have actually labelled them as disastrous.

The reforms were not an experiment. They were simply a return to orthodox economics and the old-fashioned idea of balancing budgets. The really disastrous experiment in New Zealand economic history was the creation and expansion of a 'cradle to the grave' welfare system and the protection of uneconomic industries from competition in the mistaken belief that this would create employment.

When the implementation of this experiment had squeezed the life out of the export-based farming industry, farming subsidies were added to the beneficiary list. The whole scheme finally collapsed when what was taken from it far outweighed what went into it.

Providing leadership in this period was like balancing on a knife-edge. Politicians had spent many years handing out favours to an electorate that had come to expect them as rights. Two Ministers of Agriculture, Colin Moyle for Labour and John Falloon for National, were in office for most of the period of change, although Jim Sutton spent a short time in office after Colin Moyle had announced his retirement from politics in 1990.

Colin Moyle had been a popular Minister of Agriculture in the Labour Government of 1972-75. This was a time of reasonable affluence, and he is credited with fighting hard for farmers in a Cabinet which was not known for understanding agriculture. His second term coincided with the withdrawal of all farm subsidies, and his role was largely announcing bad news. He could not have relished this, but he did not flinch from it, either.

John Falloon had been a Minister in the previous National Government, and was thus involved in the subsidy expansion. His second term came in a very different era. At first he seemed to be looking for rabbits to pull out of his hat, but found eventually that there were none.

While Moyle, Sutton and Falloon could not hand out much good news, it is to their credit that they all fought hard for balanced reform. They were especially strong on the need for tariff reduction on imports. In the new order, farmers' ability to make a living is much more dependent on overall economic management than it is on a minister of agriculture.

It has not been an easy time to be involved in politics in New Zealand, and it has not been a bed of roses for the farming leadership either. While the leaders who developed the new policies had majority support, there were plenty of critics also. Some of the criticism was bitter and personal. There were calls for direct militant action from some quarters. Such action can be very tempting for a leader to take, as it gives the impression that something is actually being done. There were marches and demonstrations, but with minor exceptions they never got out of hand. This was most important because the lines of communication must always be kept open when leaders are negotiating with politicians, bankers and other business leaders.

Making a lot of noise in public may appear impressive, but it is useless if the people you are trying to convince then refuse to speak to you. This was the experience of the leaders of some other sectors and they were of little value to their membership as a result. I am proud to say that this never happened to Federated Farmers.

Farmers are still very hard taskmasters, however. Four Federated Farmers vice-presidents, Keith Hanning, Malcolm Lumsden, Bill Shepherd and John Boddy, have all been defeated in elections in the reform period. All of them had worked hard and well but, with farm politics in a volatile state, they became scapegoats.

Rob McLagan has been a wonderful servant of the farming community over the years of change. He was Chief Executive of Federated Farmers throughout most of this time, and had considerable input into the new policies. He also was unflappable in the face of angry farmers at meetings. An irate farmer once approached him and was loudly critical of what the farming leadership was doing. Rob asked him what he would do if he were the farming president. 'The first thing I'd do,' roared the farmer, 'would be to get rid of you!'

'Oh, yes,' said Rob politely, 'and what would be the second thing you'd do?' Our farmer's face was a study. He had never thought past the notion that changing the leadership would solve all the problems.

We have been very fortunate in New Zealand that we have had a group of people in farming and political leadership for whom getting re-elected was not the major goal. Getting the country and their industry on a firm foundation was much more important.

It was not without personal cost, especially for the farming leaders. Critics often complain that leaders do not have to take the consequences of their actions if they get it wrong. The perception is that those at the top continue to draw large salaries irrespective of performance.

This was certainly not so for Federated Farmers leaders. Few of them received allowances that even covered their expenses. They were all working farmers and every decision they took was likely to impact on their own livelihood. Some of them were very hard-hit financially and one or two suffered personally because of the time spent in helping others through the crisis.

It is certainly a myth that it all came easily. Many of those who advocated and implemented change received little thanks for their efforts, but few of them would wish to undo what they did. They are proud of their part in changing the New Zealand economy and proud of the fact that we now have a well-structured, market-oriented farming industry that can compete in the world without subsidies.

The difference between the subsidy era and that prevailing today is that almost all the problem-solving, restructuring and successful development now taking place is being achieved through the skills and persistence of industry participants themselves. Although Cabinet ministers and other politicians have on occasion acted as facilitators, there is little political involvement as such, and next to no bureaucratic interference.

Many of our industries and institutions are still being refined to suit the new environment. Stripping away the subsidies and protection has made it very clear where the problems exist and where the opportunities lie.

The history of the changes in the meat processing industry provides a good example of how the farming industry has adapted. Participants in the sheep and beef cattle industries have wrestled with the search for the best processing and marketing systems for many years. Most participants have favoured a free and competitive system, but international hygiene requirements and market quotas (and at times the seasonality of the industry) have meant that some licensing has been required. It has proved difficult to get this right.

In the early days through to the 1930s, many processing companies failed. This led to an era of strict licensing in which established processors were largely protected. This protection in turn led to great inefficiency and very little innovation, and it was ultimately abandoned in 1980.

Sir Peter Elworthy, Hon Rob Storey, the Author, Sir Allan Wright, John Kneebone, and Owen Jennings — all former presidents of Federated Farmers.

The Meat Producers Board had had the power since its inception to acquire and market red meat, and it took over the marketing of sheep meat late in 1982, after the subsidy-inspired increase in production had caused a big drop in prices. The single-desk selling made little progress in a very difficult situation, however, and a significant amount of good meat proved unsaleable. Marketing was handed back to the companies again in 1985.

The delicensing of the processing industry and the handing back of sheep meat marketing to the industry saw some new people enter the business, and intense competition between the companies for farmers' stock. This competition put prices up for farmers in the short term, but it created problems too.

The first was that a number of companies began to fail again, leaving creditors (including farmers) owed a great deal of money. A farmer selling a large consignment of cattle, representing the whole year's income, was especially vulnerable in this environment if the processing company went bankrupt before it had paid him for the stock.

A second problem was the fact that while companies were spending furiously to gain throughput for their processing plants, some were not investing nearly as much as they should have been in market development and research.

There was a third problem in this highly competitive situation, especially in developing markets. I believe that the future for lamb, and to some extent beef, lies in the high quality chilled form, but while there was more money in selling frozen lamb and beef into the commodity trade, then that is what companies did. They then had the money to compete for stock in the short term. However, without product and market development it was inevitable that the industry declined and farmers were ultimately losers.

In addition to these problems, several companies laboured under very high debt loadings from a previous takeover. When an industry's raw materials (in this case, sheep) drop from 70 million to 50 million units in a very short period, some rationalisation of processing capacity is obviously necessary. This is not a straightforward operation under a system of competitive ownership. If one company owns three plants and closes one down, you might expect the throughput in the two remaining plants to improve markedly. In practice, this will not necessarily happen, as some suppliers will not send their stock to the remaining plants of the same company, but will switch to a competing plant.

The company doing the rationalising will carry all the cost, but all the others will benefit to some extent. This is why rationalisation has not taken place as rapidly as it should have done.

These problems may seem insurmountable, but they are not. I am happy to say that the participants in the industry are now by and large working together constructively to solve them. For example, when Weddel failed in 1994, most of the meat companies in the North Island joined a consortium which bought the Weddel plants from the receiver and closed them down. This meant that a very cost-effective rationalisation was carried out with all companies sharing the costs and the benefits.

The main problem this consortium had was getting approval from the Commerce Commission, New Zealand's competition watchdog. It is debatable whether, in the new environment, New Zealand's competition laws are still appropriate.

The establishment of a Meat Industry Board on which both farmers and meat companies are represented is a step in the right direction. Participants should be working together from now on, rather than against each other, as happened too often in the past.

Another very positive development has been the successful fund-raising and share floating carried out by two major meat companies, AFFCO and Alliance. Those managing these two companies have done a magnificent job of restoring the confidence of the investing public in their businesses after an extremely turbulent period.

There are some real success stories in farming development in recent years. The dairy, deer, wine and some horticultural industries have all made great progress, and the people involved in these thriving sectors deserve great credit.

On the farms, most farmers are in a sound financial position. Some are doing very well, while others still struggle somewhat. However, there are willing buyers for most farms whose owners are struggling and wish to sell out.

After years of blaming meat companies, farming leaders and banks for their poor returns, farmers are realising that there have been far greater efficiency gains made in competing meats and fibres than have recently been made with sheep meat, wool and beef. Despite these greater gains, those farmers who are applying available technology are making good returns, even with the current depressed markets.

A new attitude, that of getting on with the job and meeting the competition, is emerging. The attitude that existing products and techniques are good enough is at last dying, at least in New Zealand.

I remember hearing a radio interview early in 1993 where a British apple grower was asked if he planned to try growing any of the new apple varieties. He replied that he did not, as 'everyone knows that British Cox's Orange are the best apples in the world'. Growers with this attitude in New Zealand would soon be out of business. The aim of the New Zealand farming industry is to be the best in the world at what it does. A huge effort is being made by people in all walks of life to make this happen.

For Federated Farmers, whose members in many respects started the whole process of change, the job is still not finished. Managing all facets of the economy so that efficient export industries can operate profitably is a very exacting science, and New Zealand may not yet have it quite right. President of Federated Farmers Graham Robertson and his colleagues continue to suggest refinements. These may not all work, but the organisation is doing its job in keeping the ideas coming and having them thoroughly tested and debated.

The best part of farming without subsidies is that we are all free to choose our own path in farming. We stand or fall on our own abilities and choices, with little bureaucratic interference and no government intervention. We would not go back to subsidies again - why would we want to?

Imagine a world where the Uruguay Round of the GATT was just the starting point in agricultural trade reform. Imagine the wealthy countries farming to comparative advantage, exporting any surpluses without subsidy and importing other products from countries that can grow them more efficiently.

Imagine the dismantling of farm subsidy programmes worldwide, and the money saved being spent on helping the Third World to grow its own food and feed all its people.

Imagine the world's consumers with more choices and fewer expenses in buying their food each week. Imagine the reductions in taxation there could be in country after country, and how that money could be used.

Imagine what could happen in a world where those who have the power to make a difference stopped believing the myths about farming and subsidies.

THE AUTHOR

Brian Chamberlin has had a varied career - farmer, agricultural journalist, farmer politician, broadcaster, lobbyist, diplomat and consultant on agriculture and trade. The thread running through all of this, however, has been his first love, which is unquestionably farming.

Brian Chamberlin and his wife and business partner Jan bought a run-down 340-acre sheep and cattle farm in 1967 on a deposit of $NZ8,000. Their farming company, Euroa Farms Limited, has been expanded over the years, and they now farm 1,100 acres (440 hectares) freehold and 800 acres (320 hectares) leasehold, in two separate locations.

The 1967 farm purchase coincided with one of the downturns in New Zealand agriculture. The struggle to service a large mortgage and make ends meet was a huge challenge for the young couple. In addition to running the farm and raising two children, they built up an agricultural contracting business, specialising in shearing and haymaking, to bring in some much needed income. Later off-farm work included writing for local papers, editing a monthly farming newspaper, and hosting a top rating farm-related radio programme.

The difficult situation in farming eventually led Brian into agricultural politics and commerce, and Jan into farm tourism and off-farm work.

Brian Chamberlin held many positions in the New Zealand farming organisation, Federated Farmers, including Auckland Provincial President and National Meat and Wool Chairman. He was a National Vice-President for six years before becoming National President from 1987-90.

At about the same time, he became a Vice-President of the International Federation of Agricultural Producers, a position he held for four years.

His leadership role in Federated Farmers at national level coincided with both the era of development of subsidies in New Zealand agriculture and the era of their dismantling. He chaired a Federated Farmers committee in 1982 which called for inflation control, rather than increased subsidies. In 1984 he was the Federated Farmers representative on the committee which organised the Economic Summit Conference.

The period of Brian Chamberlin's presidency of Federated Farmers also coincided with that of the Uruguay Round of the General Agreement on Tariffs and Trade, and with agriculture on the agenda for the first time in a GATT round, he became increasingly involved in lobbying for a successful conclusion.

He travelled to many countries as President of Federated Farmers, and at the end of his term as President, was appointed Special Agricultural Trade Envoy for New Zealand by the then Labour Trade Minister, the Right Hon. Mike Moore. This was a position which involved carrying New Zealand's case for trade reform around the world, and he was subsequently reappointed to the position by the new National Government's Minister for Trade Negotiations, the Hon. Philip Burdon, in 1991.

In 1992, he was appointed Counsellor (Agriculture) to the New Zealand High Commission in London, to continue the work.

In 1993 he returned to New Zealand, where he has since chaired a review of Tradenz and been involved in the restructuring of the kiwifruit and meat processing industries.

He represents New Zealand on the International Policy Council for Agriculture and Trade, is on the Australian Advisory Council for SES Ltd, and is an advisor to the Pacific Islands radio station in Auckland, 531 pi.

Past positions include directorships of a number of farm co-operative companies, and chairmanship of Radio Pacific Ltd.

While still remaining their business interests in New Zealand, Brian and Jan Chamberlin recently moved to Australia, to see more of their son and daughter and five small grandsons.

APPENDIX

Figure 1:
TOTAL FERTILISER SALES.
Fertiliser sales from the Meat and Wool Boards' Annual Review of the Sheep and Beef Industry (based on information from MAF).

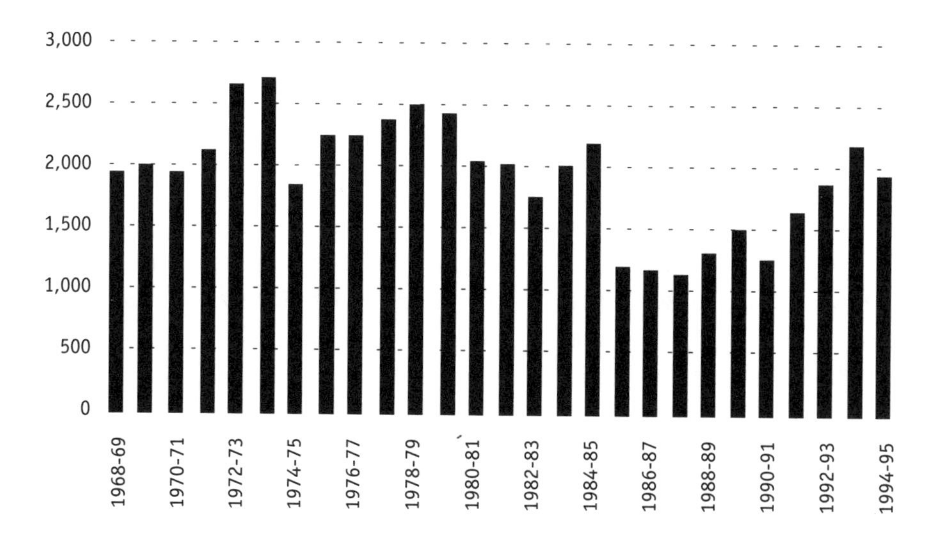

Figure 2:

LIVESTOCK NUMBERS 1960-1 TO 1994-5.

(Source: Meat and Wool Boards Economic Service using information from Statistics New Zealand)

	SHEEP (a)		BEEF CATTLE (b)		DAIRY CATTLE (b)			
June Year	Breeding Ewes	Total Sheep	Breeding Cows	Total Beef Cattle	Dairy Cows and Heifers in Calf or Milk	Total Dairy Cattle	Total Deer	Total Goats
1960-61	32,633	47,134	1,047	3,334	1,929	3,111		
1965-66	37,178	53,748'	1,214	3,856	2,088	3,362		
1970-71	42,912	60,276	1,557	5,280	2,239	3,539		
1975-76	41,108	55,320	2,311	6,294	2,387	2,998		
1978-79	44,515	62,163	1,916	5,507	2,300	2,911		
1979-80	46,108	63,523	1,823	5,122	2,287	2,900	42	49
1980-81	48,245	68,772	1,879	5,162	2,293	2,969	104	53
1981-82	49,349	69,884	1,782	5,113	2,271	2,922	109	68
1982-83	50,810	70,301	1,576	4,906	2,313	3,007	151	93
1983-84	50,966	70,263	1,448	4,497	2,404	3,134	196	150
1984-85	51,182	69,739	1,440	4,531	2,483	3,246	259	230
1985-86	50,187	67,854	1,456	4,613	2,546	3,308	320	427
1986-87	47,491	67,470	1,498	4,881	2,622	3,398	392	723
1987-88	45,382	64,244	1,586	4,804	2,551	3,195	500	1,054
1988-89	44,041	64,600	1,460	4,858	2,562	3,200	606	1,301
1989-90	41,414	60,569	1,355	4,526	2,621	3,302	780	1,222
1990-91	40,453	57,852	1,386	4,593	2,723	3,441	976	1,063
1991-92	36,631	55,162	1,388	4,671	2,642	3,429	1,130	793
1992-93	36,684	52,568	1,419	4,676	2,723	3,468	1,135	533
1993-94	35,375	50,298	1,463	4,758	2,808	3,550	1,078	353
1994-95e	34,038	49,014	1,576	5,013	2,890	3,664	1,238	284

Figure 3:

NOMINAL FARMGATE OUTPUT PRICES AND INPUT PRICE INDICES.

Significant years to note are 1976 (the base year), a year when subsidies were increased; 1985/6 when the impact of subsidy removal was really felt; and 1989, when the benefits of reform were starting to flow into prices. Reprinted from *Farming Without Subsidies*, MAF 1990.

June Year	Wool (c/kg)	Lamb (c/kg)	Mutton (c/kg)	Sheep Input Prices	Beef (c/kg)	Cattle Input Prices	Milkfat (c/kg)	Dairy Input Prices
1965	73	44	18	52	27	58	77	49
1966	72	41	18	54	37	53	81	50
1967	60	33	16	55	33	55	81	51
1968	45	40	16	57	42	57	75	58
1969	56	49	17	58	45	58	71	60
1970	51	44	23	60	59	60	72	62
1971	48	42	17	62	55	62	86	64
1972	60	38	14	66	51	60	121	68
1973	136	68	53	69	73	70	116	72
1974	132	70	41	79	48	79	128	81
1975	84	47	15	89	31	89	127	90
1976	146	73	31	100	54	100	141	100
1977	199	98	53	118	60	119	153	118
1978	169	92	44	136	66	137	167	135
1979	197	109	52	148	119	149	173	147
1980	242	1 18	58	180	122	183	208	179
1981	225	124	63	222	124	227	260	219
1982	293	164	55	254	148	261	300	251
1983	293	168	66	285	162	292	360	283
1984	291	176	76	287	169	295	350	286
1985	338	191	92	322	236	331	396	317
1986	313	108	22	362	165	373	400	358
1987	370	65	58	389	16~	403	355	388
1988	404	124	56	412	172	428	407	412
1989	458	146	54	433	222	450	5/0	435

Prices of farm inputs, weighed by expenditure

Figure 4:

RURAL SALES ACTIVITY.

Note the increases in 1980, 81 and 82, which were years of high subsidies. Compiled by the Meat and Wool Boards' Economic Services, using information from Valuation New Zealand.

Year	Number of Sales	Average Area(ha)	Price per Hectare $	Farm Land Price Index	% Change from Previous Year
1970	4210	71	376	103	+4.1
1980	4725	97	1395	463	+23.2
1981	5230	88	2008	708	+40.2
1982	3774	73	2941	932	+31.6
1983	2515	63	3128	928	-0.4
1984	3077	75	2957	969	+4.4
1985	2587	69	3085	967	-0.2
1986	1928	65	2793	933	-3.5
1987	2748	88	2462	921	-1.3
1988	2843	85	2390	882	-4.2
1989	4245	98	2372	1000	+13.4
1990	4286	104	2548	1188	+18.8
1991	3485	92	2547	1181	-0.6
1992	4343	99	2901	1358	+15.0
1993	4056	109	3215	1696	+24.9
1994	3609	99	4389	2073	+22.2

Source: Valuation New Zealand.

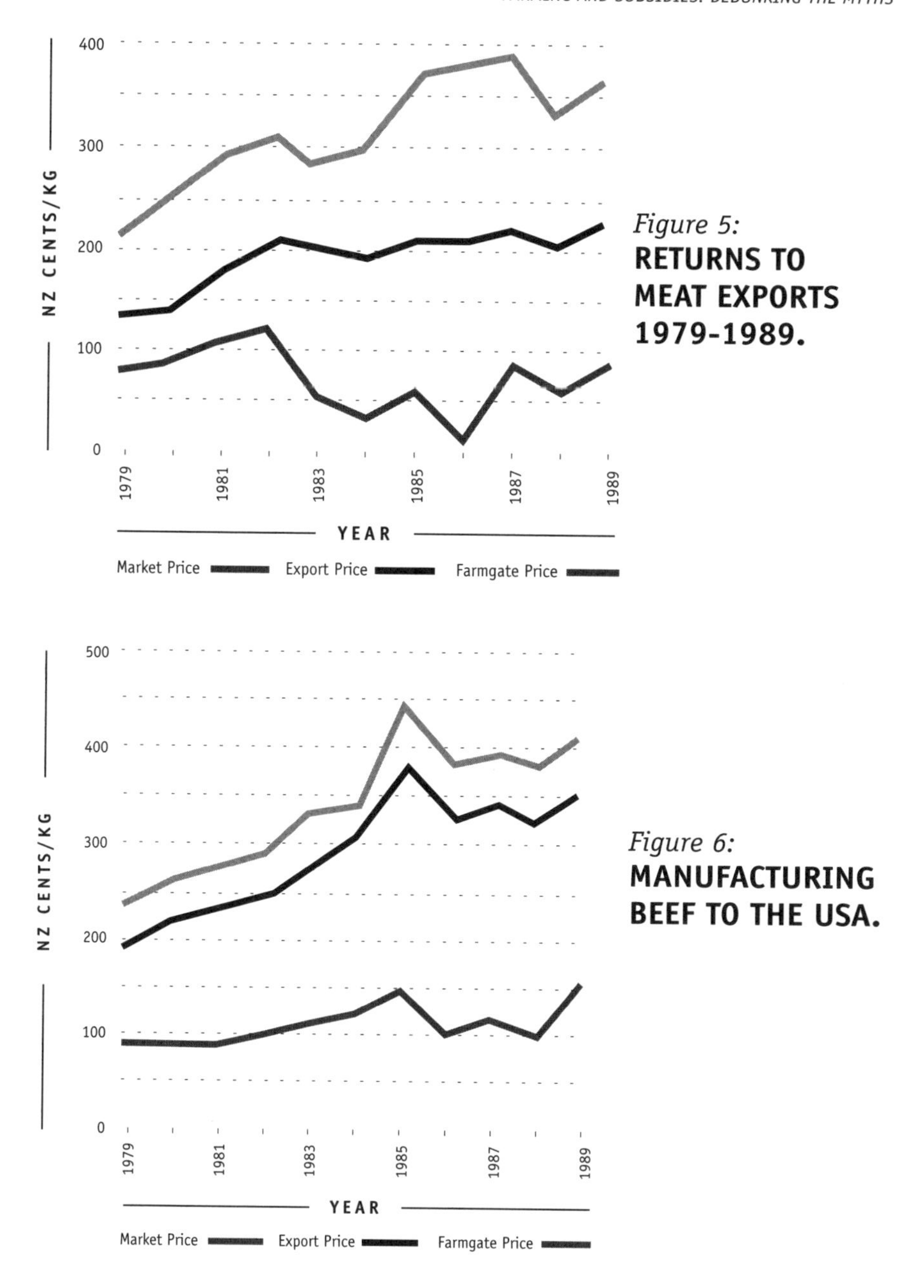

Figure 5:
RETURNS TO MEAT EXPORTS 1979-1989.

Figure 6:
MANUFACTURING BEEF TO THE USA.

These figures reprinted from *Farming Without Subsidies* (MAF, 1990) demonstrate that the market price increased much more than that paid to farmers because of cost increases beyond the farmgate.

Figure 7:

EMPLOYMENT IN AGRICULTURE AS A PERCENTAGE OF CIVILIAN EMPLOYMENT.

Reprinted from SONZA (Situation and Outlook New Zealand Agriculture), 1995, MAF.

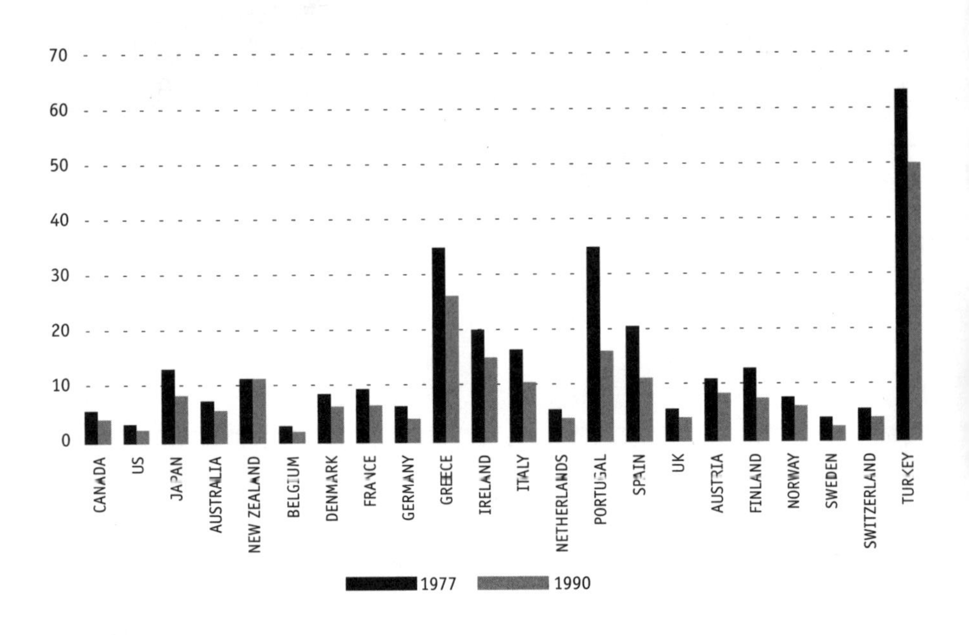